THE FORMATION OF SOIL MATERIAL

The Formation of Soil Material

T. R. PATON

Senior Lecturer, Earth Sciences,
Macquarie University, Sydney, Australia

London
GEORGE ALLEN & UNWIN
Boston   Sydney

First published in 1978

GEORGE ALLEN & UNWIN LTD
40 Museum Street, London WC1A 1LU

© T. R. Paton, 1978

British Library Cataloguing in Publication Data

Paton, T R
   The formation of soil material.
   1. Soil formation
   I. Title
   631.4   S592.2

   ISBN 0–04–631009–6   Papercase
   ISBN 0–04–631010–X   Paperback

Typeset in 10 on 12 point Times by
Bedford Typesetters Ltd
and printed in Great Britain
by Biddles Ltd, Guildford, Surrey

To the memory of
Edward Crompton

Preface

My fifteen years of soil survey work in Malaysia, Africa and Australia resulted in a feeling of general unease regarding the conceptual foundations on which such work was based. Since becoming a university teacher ten years ago, I have had the opportunity to examine the foundations of pedology with a view to clarifying these concepts. I soon realised that a great number of problems were generated as a result of illogical and restrictive thinking on many basic issues. Analysis and rationalisation of these forms the basis of this book which as a result presents a model of soil material formation differing considerably from those which have been advanced previously.

The sequence of topics in the book is dictated by the way in which the model is constructed and may be divided into four parts: Chapters 1 to 5; Chapter 6; Chapters 7 and 8; and Chapters 9 and 10. The first five chapters are concerned with the transformation and rearrangement of minerals consequent upon their being exposed to conditions at or near the earth's surface. Chapter 1 deals with the initial state of the minerals, while Chapter 2 is concerned with their breakdown (weathering) and Chapter 3 considers the differential movement of the breakdown products (leaching). The formation of new minerals and new fabrics is considered in Chapters 4 and 5 respectively. The second part consists of Chapter 6 only, in which the impact of processes of lateral surface movement is evaluated. Chapters 7 and 8, which make up the third part, integrate the biospheric reactions into the model. Insofar as it is possible to separate them clearly, Chapter 7 considers those reactions of the biosphere with the processes described in the first part and Chapter 8 with those described in the second. Up to this point only processes have been dealt with. Chapter 9 is concerned with the relationship of these processes to the factors of soil formation, while Chapter 10 develops the idea of pedological provinces arising from the previous analysis as a method of classifying soils on a world scale. This method is fundamentally different from previous approaches to the subject of soil classification.

It should be stressed that the bibliography is highly selective, listing only papers and books mentioned in the text. In particular, no reference has been made to the great amount of Russian literature, for, without personal knowledge of Russian soils, gross errors in the interpretation of Russian work could occur.

Many colleagues at Macquarie University have contributed to this book in a variety of ways. In particular I am deeply grateful for the critical appraisal of the manuscript by Mr M. F. Clarke and the help given in evaluating field problems by Mr P. B. Mitchell. In addition, considerable portions of the manuscript have been read and commented upon by Dr M. A. J. Williams, Professor J. L. Davies and Dr D. A. Adamson and to them I owe a great deal of thanks. The final critical evaluation of the manuscript by Professor A. Young (University of East Anglia), Dr A. Warren (University College, London) and Professor D. Branagan (University of Sydney) is particularly appreciated.

Mrs A. Coates was responsible for preparing the typescript, together with Mrs C. Gibson, Mrs O. Zakroczymski, Miss G. Keena and Miss L. Johnston. The figures and tables were prepared by Mr R. Bashford, Mr D. Ellis, Mr D. Oliver, Mr K. Rousell and Miss S. Knight and to all of them I am most grateful for their patience and precision.

Throughout the whole period during which the book was being written the work was supported by research grants from Macquarie University and this help is gratefully acknowledged.

A final acknowledgement must be made to the debt which I owe to the late Edward Crompton, who was responsible for the initial promptings which have eventually resulted in this book.

T. R. PATON,
Macquarie University,
School of Earth Sciences
December 1977

Acknowledgements

Cases in which my Figures and Tables are derived from the work of other scholars are indicated in the captions. In addition, I am indebted to the following for permission to reproduce four of the Figures: Cambridge University Press, for Figure 2.1 from *An introduction to crystal chemistry*, by R. C. Evans. Blackwell Scientific Publications, for Figures 8.1 and 8.2 from the *Journal of Ecology* **52,** pp. 370 and 371. Williams & Wilkins Co., Baltimore, for Figure 7.1 from *Soil Science* **60,** p. 30.

Contents

List of Tables

1

The composition of the lithosphere

It is normal, when considering the problem of soil formation, to start with a definition of soil. However, immediately this approach is taken many of the problems with which this book is concerned would be anticipated, for subsequent discussion would then be restricted automatically to conform with this definition. This dilemma can be avoided by considering initially the *processes* that are of primary interest to pedologists, rather than the *materials* that result from the processes which are the subject of most definitions. From this point of view, the pedologist's centre of interest is concerned with those processes that are going on at the surface of the land masses of the world, with a decline in interest the further any process is removed from this surface. In the case of water, for example, pedologists are very concerned with the ways in which it reacts with minerals forming part of the lithosphere, but are much less concerned with the way in which water circulates in the atmosphere. Again, in considering the minerals making up the lithosphere, the way in which they break down at the interface is of major concern to pedologists, but their interest in how these minerals were originally formed is much less. The point to note is that no absolute boundaries are being defined; it is only that interest declines away from the lithosphere–atmosphere interface. A core area is being defined with a large peripheral zone.

Composition of the lithosphere

To evaluate the formation of soil material it is necessary, therefore, to concentrate on reactions between the lithosphere on the one side and the atmosphere and biosphere on the other. In particular this means an understanding of the volume relationships between a given mass of soil material and the lithospheric constituents from which it has been derived. Initially, this requires some knowledge of the composition of that part of the lithosphere forming the land areas of the Earth. These are, in some respects, remarkably heterogeneous, being formed of a mosaic of sediments, metamorphosed sediments, igneous intrusives and volcanics, faulted and broken into blocks of various shapes and sizes. However, in terms of elemental composition 99% of lithospheric material consists of only eight elements: oxygen, silicon, aluminium, iron, magnesium, calcium, sodium and potassium. The left-hand column of Table 1.1 expresses this fact in terms of weight per cent. By dividing the amount of each of these elements by its atomic weight and expressing the results as a percentage, the factor of density is eliminated. The result, given in the second column of Table 1.1 as 'atom per cent', is an expression of the relative abundance of atoms of these eight elements in the Earth's crust. By taking account of the relative volume of these atoms (Fig. 1.5), it is possible to calculate the volume occupied by each of these atoms in a standard volume of the Earth's crust, as is done in the third column of Table 1.1. From this it is apparent that oxygen accounts for almost 94% of the volume of the Earth's crust while the other seven major constituents account for just over 6 per cent.

Table 1.2 (Barth 1948a) shows average igneous rock composition in terms of volume per cent. The average basalt can be taken as being representative of oceanic crustal areas and the shield granite as representative of the continental crustal areas. In both cases oxygen is overwhelmingly dominant and the differences are mainly in the higher values for iron, magnesium and calcium in the basalt and for sodium and potassium in the granites. In terms of either weight or volume, the

Table 1.1 The commoner chemical elements in the Earth's crust (after Mason 1952).

	Weight per cent	Atom per cent	Volume per cent
O	46·60	62·55	93·77
Si	27·72	21·22	0·86
Al	8·13	6·47	0·47
Fe	5·00	1·92	0·43
Mg	2·09	1·84	0·29
Ca	3·63	1·94	1·03
Na	2·83	2·64	1·32
K	2·59	1·42	1·83
	98·59	100·00	100·00

lithosphere contains far more oxygen than is found in the atmosphere. This accumulation is made possible only by the small number of positively-charged ions which hold together the negatively-charged oxygens. The atom per cent column of Table 1.1 shows that the most abundant cation is silicon and so, to achieve neutrality, there would be a greater tendency for combination to occur between silicon and oxygen than anything else. This explains why silicates are the most common materials of the Earth's crust.

Silicate structures

To a first approximation, size controls the way in which silicon and oxygen combine (see Fig. 1.5). Geometrically it is possible to arrange four oxygen ions around this silicon so that all are touching (Fig. 1.1). It is not possible to arrange any more and, therefore, it can be said that silicon has a co-ordination number of 4 with oxygen. These oxygens are arranged symmetrically in space around the central silicon and this symmetrical arrangement is naturally that of a tetrahedron. The silicate tetrahedron is the fundamental building block of all the silicate minerals of the Earth's crust.

There is another important factor which has to be taken into account before this basic building block can be used to assemble more complex structures: the amount of charge carried by the ions. In the case being considered silicon is quadrivalent while oxygen is divalent. Therefore, in these terms, neutrality would require one silicon to combine with

Table 1.2 Chemical composition of typical rock expressed in volume percentages (after Barth 1948).

	Average basalt	Average igneous rocks	Shield granite
O	91·11	91·83	92·12
Si	0·70	0·83	0·92
Ti	0·12	0·05	0·02
Al	0·74	0·79	0·76
Fe	1·47	0·58	0·21
Mg	1·09	0·58	0·09
Ca	2·78	1·50	0·45
Na	1·28	1·68	1·75
K	0·70	2·19	3·68
	99·99	99·99	99·99

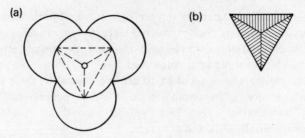

Figure 1.1   (a) The basic tetrahedron of silicon and oxygen. (b) Conventional representation of tetrahedron structure.

two oxygens, that is SiO_2, whereas in terms of geometry one silicon is required to combine with four oxygens, that is SiO_4, which would carry a negative charge of 4. Silicate structures can be explained in terms of how the geometric and valency constraints are reconciled. There are two ways in which this is possible: first, by joining together tetrahedra so that oxygens are shared between neighbouring silicons, thus reducing the negative charge deficit; secondly, by making use of the positive charges of other metal cations to balance the negative charge. Both of course occur together to produce a neutral mineral entity but at this point it will be better to consider them separately.

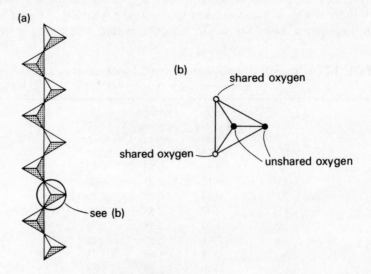

Figure 1.2   (a) Inosilicate single-chain tetrahedra linked by corners. (b) Details of linking in one tetrahedron.

Tetrahedral links The simplest way to link tetrahedra is merely to join one corner with another as a single chain, as shown in Figure 1.2a. All the silicons in these tetrahedra are naturally in four co-ordination with oxygen and the way in which the charge distribution has altered from the state of isolated tetrahedra is shown in Figure 1.2b. Two of the oxygens in any one tetrahedral unit are shared with neighbouring tetrahedra while two are not. This reduces the valency contribution of the shared oxygens from two to one; therefore, the negative charge provided by the oxygen in any one tetrahedral unit of the chain is six, as opposed to eight for isolated tetrahedra. This means, in turn, that the negative charge imbalance is reduced to 2 from 4 if the length of the chain is considered to be infinite.

A more complex configuration is shown in Figure 1.3, in which two single chains are cross-linked. Again ignoring the terminations of the chain, it is possible to distinguish two types of tetrahedra: external tetrahedra (Fig. 1.3b) and internal tetrahedra (Fig. 1.3c). The external tetrahedra are linked to two adjoining tetrahedra as in the case of the single chain; it is exactly the same situation as shown in Figure 1.2b and so for these tetrahedra the imbalance of negative charge is 2. However, the internal tetrahedra of Figure 1.3c now share three oxygens with adjoining tetrahedra and have only one oxygen, the apical one, not so linked. Therefore, these tetrahedra have one oxygen and three half oxygens, which gives a total negative charge of 5 as against

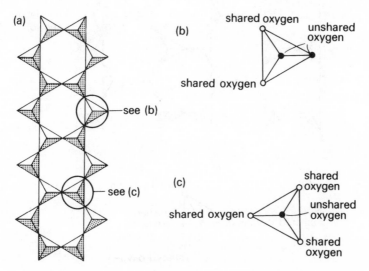

Figure 1.3 (a) Inosilicate double-chain cross-linking between alternate tetrahedra. (b) Details of external tetrahedra. (c) Details of internal tetrahedra.

the positive charge of 4 for the silicon. There are equal numbers of these two types of tetrahedra in the double chain structure; therefore, the overall negative imbalance for each silicon is $1\frac{1}{2}$ as compared with 2 for the single chain and 4 for isolated tetrahedra. Increased linking has decreased the overall charge imbalance.

The natural progression from this stage is to the sheet structure (Fig. 1.4) in which all tetrahedra are sharing three oxygens with neighbouring tetrahedra. Therefore, all tetrahedra away from sheet edges are as shown in Figure 1.4b, which is exactly the same as Figure 1.3c, which makes the overall negative charge imbalance 1, reduction again accompanying increased linkage.

The maximum linkage is attained in a three-dimensional network in

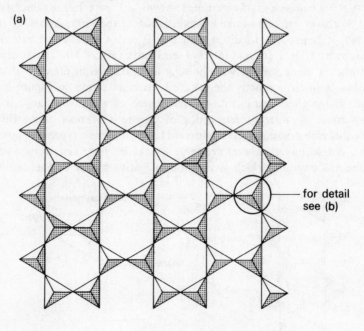

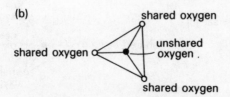

Figure 1.4 (a) Phyllosilicate sheet structure arising from linking of three neighbouring tetrahedra. (b) Details of tetrahedral linking.

which all oxygens are shared between tetrahedra; each silicon has associated with it four half oxygens. This gives rise to four positive and four negative charges for any tetrahedral unit, thereby achieving an electrically balanced structure. Thus, by appropriately linked tetrahedra it is possible to reconcile the disparate geometric and valency controls on silicate structure.

Those minerals that are made up of independent silicate tetrahedra are classified as **orthosilicates,** typical of which are the olivines. Minerals with chain structures as in Figures 1.2 and 1.3 are classified as **inosilicates**: an example of the single chain inosilicates is provided by the pyroxenes, whereas the double chain is represented by the amphiboles. The sheet structures, as shown in Figure 1.4, are the **phyllosilicates,** of which the micas are a typical example. Minerals made up of continuous three-dimensional networks are called **tektosilicates** and the most outstanding example of this type is quartz.

Cation content All of these minerals, such as olivines, pyroxenes, amphiboles, micas and quartz, are electrically neutral. Therefore, except for quartz, all must accommodate other cations in addition to silicon to achieve this neutrality. There are two aspects to be considered: the total amount of cations needed and the kind of cations required.

In considering the quantity of cations needed, it is necessary to take a constant volume of material having each of these silicate structures in turn – i.e. orthosilicate, inosilicate (single and double chain), phyllosilicate and tektosilicate – and to calculate the total quantity of positive charge required by this volume to achieve neutrality. Due to the relatively large size of the oxygen, the lithosphere can be considered as a packing of oxygen anions with cations in the gaps between them. By basing calculations on a standard number of oxygen ions, standard volumes of silicate minerals are, in effect, being dealt with (Barth 1948b). In the following discussion, 100 oxygen ions are taken as a basis for calculation. In the case of orthosilicates, each isolated tetrahedron has the formula SiO_4. Therefore, for the four oxygens of this basic unit there are four negative charges and for the standard volume of 100 oxygens there will be 100 negative charges.

In the case of the single chain inosilicates, of the four oxygens in a single tetrahedral unit two are shared with neighbouring tetrahedra. Therefore, only half of each of them can be considered as belonging to the single tetrahedral unit. These two halves, together with the two unshared oxygens, give a formula of SiO_3, signifying that there is a negative charge of two for every three oxygens; for a standard volume of 100 oxygens there will be $66\frac{2}{3}$ negative charges. Double-chain inosilicates are rather more complex because, as Figure 1.3 shows, two different types of tetrahedra are being dealt with. Half of them share two oxygens with neighbouring tetrahedra, as for single chain inosilicates, giving a formula of SiO_3. The other half shares three oxygens with neighbouring tetrahedra (Fig. 1.3) which gives them a formula of $SiO_{2\frac{1}{2}}$. The average value for double chain inosilicates is therefore, $SiO_{2\frac{3}{4}}$. Consequently, for every $2\frac{3}{4}$ oxygens there is an excess negative charge of $1\frac{1}{2}$: for a standard volume of 100 oxygens the negative charge would be $54\frac{1}{2}$. A single tetrahedron in the phyllosilicate structure shares three oxygens with neighbouring tetrahedra and has an overall formula of $SiO_{2\frac{1}{2}}$. For every two-and-a-half oxygens there is one negative charge; for a standard volume of 100 oxygens there are 40 negative charges.

In the case of tektosilicates all four oxygens in any given tetrahedron are shared with neighbouring tetrahedra and the formula is SiO_2. The four positive charges of the silicon are balanced by the four negative charges of the oxygens; in a standard volume of 100 oxygens there would be no excess negative charges.

Therefore it can be seen (Table 1.3) that the number of metal cations per unit volume required by these various silicate structures to achieve neutrality will be most abundant in the orthosilicates and will decrease towards the tektosilicates; that is, the more complex the silicate framework, the fewer cations involved (other than silicon).

Now consider the kinds of cations which may occur in these various structures. A major control is their size (Fig. 1.5). In general, the greater the charge, the smaller the size. Potassium with a single charge (univalent) approaches an oxygen in size, whilst trivalent aluminium is relatively small; divalent ferrous iron is larger than the trivalent ferric form. It is obvious that the structure would require large holes to accommodate ions such as potassium, calcium and sodium, but much smaller holes would suffice for magnesium and iron. In general, the more tetrahedral linkages there are in any particular structure, the larger the spaces that are developed. The more this Si–O–Si bond is

Table 1.3 Number of positive charges required by 100 oxygens to attain neutrality in various silicate structures.

Orthosilicates	100
Inosilicates (single chain)	66
Inosilicates (double chain)	54
Phyllosilicate	40
Tektosilicate	0

developed, the more it resists close packing; thus orthosilicates can be very tightly packed, inosilicates are less capable of being closely packed and tektosilicates generally have very open structures. The holes in orthosilicate structures are very much smaller than those in the tekto-silicates; therefore, smaller metal ions, such as iron and magnesium, would be accommodated in the former and potassium, calcium and sodium could be preferentially accommodated in the latter.

Electronegativity   The tendency for particular cations to be associated with particular structures is reinforced by the type of bond developed. Two main types of bonding can be distinguished: the ionic, in which electrons are transferred from one atom to another to produce a negative and a positive ion; and the covalent, in which, instead of electrons being transferred from one atom to another, they are shared equally between the two atoms. In most cases, however, it is not an either/or affair, but one in which the bonds partake of both ionic and covalent characteristics. This tendency is expressed in the concept of **electronegativity**: that is, the power of an atom in a molecule to attract electrons to itself. When two atoms of similar electronegativity are joined by a bond it will be covalent, since both atoms tend to attract the bonding electrons to the same extent and hence the electron distribution is symmetrical. If, however, one of the atoms has a higher electronegativity than the other, it will draw the shared electrons closer to itself, whereby an ionic component is introduced into the bond. Electronegativities for the eight most abundant elements are given in Table 1.4. The greater the difference between the electronegativities, the more ionic the bond between these elements; the closer the figures, the more covalent will be the bond. In silicate minerals the Si–O–Si bond is the most common; it has been found that the greater the number of such bonds in a particular mineral, the more the electronegativity of oxygen is increased. That is, the effective electronegativity

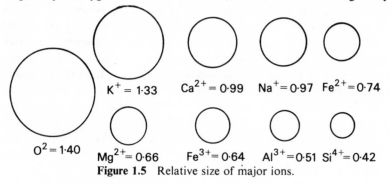

$K^+ = 1.33$ $Ca^{2+} = 0.99$ $Na^+ = 0.97$ $Fe^{2+} = 0.74$

$O^2 = 1.40$ $Mg^{2+} = 0.66$ $Fe^{3+} = 0.64$ $Al^{3+} = 0.51$ $Si^{4+} = 0.42$

Figure 1.5   Relative size of major ions.

Table 1.4 The electronegativities of the major silicate elements.

Na	K	Mg	Ca	Fe	Al	Si	O
0·9	0·8	1·2	1·0	1·8	1·5	1·8	3·5

value of oxygen increases stepwise through the sequence: orthosilicate, inosilicate, phyllosilicate, tektosilicate. Those cations of low electronegativity values, such as sodium and potassium, will be preferentially included in the tektosilicates where the oxygen is more electronegative, but iron will be preferentially included in the orthosilicates where the oxygen is less electronegative.

Ion interchangeability It has been found that, if ions do not differ in size by more than 15% and in valency by more than one unit, they can substitute for one another in the silicate minerals. Thus, from Figure 1.5 it can be seen that sodium and calcium can replace one another completely, whereas potassium is too large and cannot react in the same way; this is despite the fact that calcium is divalent and sodium monovalent. In the same way, magnesium and ferrous iron can replace one another completely, both being divalent and within 15% of one another in size.

Aluminium must be dealt with separately because of its great importance. It is intermediate in size and valency between silicon on the one side and magnesium and iron on the other, so that it is capable of replacing any of these. Consider aluminium replacing silicon in its tetrahedral position. According to Table 1.4 silicon has a higher electronegativity than aluminium, 1·8 as opposed to 1·5. As already mentioned, the electronegativity of the oxygen in the silicate structures increases as the degree of bonding increases from orthosilicates to tektosilicates so that, in the same direction, there is an increasing tendency for aluminium, with its lower electronegativity, to be preferred to silicon in the tetrahedral position. Thus, in the tektosilicates, up to one quarter of the tetrahedral silicon can be replaced by aluminium. However, aluminium is only trivalent whereas silicon is quadrivalent; the substitution leads to an imbalance of negative charge in these tektosilicates. This is balanced by the inclusion of cations of low electronegativity and large size such as potassium, sodium and calcium, which gives rise to the feldspars.

To show how all these different structural factors are combined in a mineral an example from each of the main silicate groups will be discussed in somewhat greater detail.

Orthosilicates Figure 1.6 is a representation of the structure of olivine, a typical orthosilicate. It consists of individual SiO_4 tetrahedra. Close packing is achieved by having the tetrahedra alternately pointing in opposite directions, for without inter-tetrahedral bonding there is no stability to preserve an open structure. The tetrahedra are bound to one another and the negative charge of the SiO_4 units balanced by magnesium and/or iron. This close-packed structure of the tetrahedra is reflected in the high density of olivine compared with other silicates. There is little replacement by other elements: in particular, there is no substitution by aluminium of the tetrahedral silicon or the inter-tetrahedral iron or magnesium.

Inosilicates The single chain inosilicate is represented by the pyroxenes (Fig. 1.7a). An end-on view of the chain (Fig. 1.7b) shows that the individual chain units can be represented by a trapezium shape. Figure 1.7c shows the way in which these trapezium-shaped single chains are arranged within a pyroxene crystal. The positions occupied by non-framework cations between the chains are also indicated. These cations can be ferrous and ferric iron, magnesium, calcium and aluminium. In addition there is a certain amount of substitution of silicon by aluminium within the chains. In consequence the composition of pyroxenes is extremely variable.

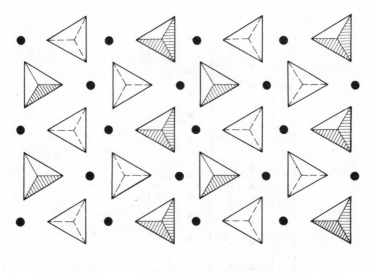

● Fe+Mg

Figure 1.6 Olivine.

Amphiboles, as examples of double-chain inosilicates, are shown in Figure 1.8; the only difference to note from the previous diagram is that the double chain, when viewed end on (Fig. 1.8b), is represented as a double trapezium unit; stacking them together in the crystal (Fig. 1.8c) gives rise to two different types of hole in the structure: (a) between the unshared apical oxygens and (b) between the shared basal oxygens. Into the first of these holes is fitted a wide range of cations including iron, magnesium, calcium and aluminium, as well as the much larger hydroxyl ion, which fits into the gap within the double trapezium unit.

Into the second type of gap fits the larger potassium ion with its single charge. Again, as in the case of the pyroxenes, a certain amount of silicon within the chain units is replaced by aluminium. The composition of amphiboles is thus even more complex than that of pyroxenes.

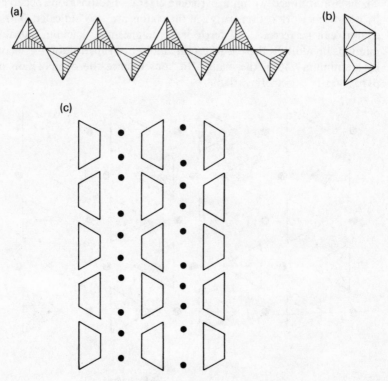

Figure 1.7 Pyroxenes: (a) Chain. (b) End-on view of chain. (c) **Packing of** chains.

Phyllosilicates The phyllosilicate structure, as exemplified by the micas, can be regarded, for purposes of description, as a rationalisation of the amphibole structure. Instead of double trapezium units facing alternate ways in any one row, all double trapezium units in any one row face the same way and are joined to one another, as shown in the change from (a) to (b) in Figure 1.9. These trapezium sheets then occur either with apices opposed or with bases opposed in pairs, as shown in Figure 1.9c. Due to this rearrangement the spaces have also been re-arranged so that type (i) space is segregated from type (ii) space and each type is restricted to separate planes. Type (i) space is occupied by the

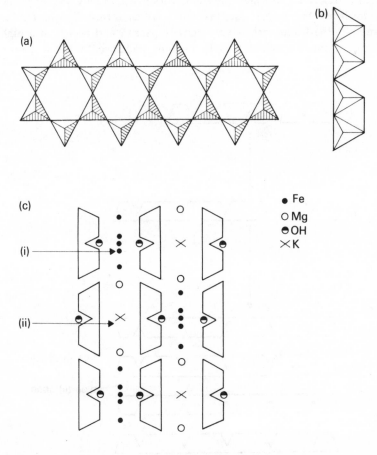

Figure 1.8 Amphiboles: (a) Double chain. (b) End-on view of chain. (c) Packing of chains.

same type of cations as in the amphiboles – that is, magnesium, iron, aluminium and hydroxyl which, due to numerous linkages, tend to hold the sheets on either side closely together. Type (ii) space is occupied only by the large and weakly bonded potassium ions, so that the sheets on either side are held only weakly. This produces the high degree of cleavage associated with micas. Again, due to the extremely complex substitution present in type (i) space, micas of very complex composition occur.

Tektosilicates In this structure all oxygens are shared between neighbouring tetrahedra. In contrast to the other types of Si–O–Si linkage already described, in which the framework exists in only one form, the tektosilicate framework can take on various forms. In quartz, for example, the framework is very densely packed and occurs in a high degree of purity which is strongly resistant to change, for the structure,

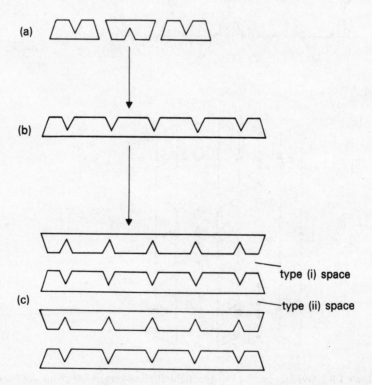

Figure 1.9 Mica: (a) End-on view of amphibole double chains. (b) End-on view of phyllosilicate sheet. (c) Packing of phyllosilicate sheets in mica.

being electrically neutral, prevents any form of substitution. In the majority of tektosilicates, however, the framework is far more open than that found in quartz; this makes them notably less dense than the orthosilicates, inosilicates and phyllosilicates. Feldspars are the most important minerals of this type. The basic structure, represented in Figure 1.10, is a ring structure made up of four tetrahedra. In two of the tetrahedra the apical oxygens point upwards whilst in the other two the oxygens point downwards. These apical oxygens provide the links

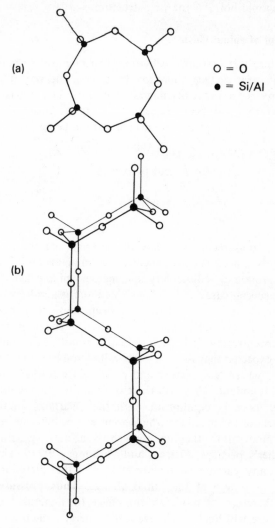

(a)

O = O
● = Si/Al

(b)

Figure 1.10   Feldspar: (a) Plan view. (b) Elevation.

in both directions with neighbouring ring structures, giving rise to a kinked chain which runs vertically through the mineral. This appearance is, however, somewhat deceptive, for it must be remembered that all the oxygens of the tetrahedra are shared, so that bonds also occur transversely between the chain structures, producing a three-dimensional network. In feldspars a quarter of the tetrahedral silicon is replaced by aluminium; the deficit of charge is made up by the accommodation of the large potassium, sodium and calcium ions in the large hexagonal holes of the ring structure.

The concept of epimorphism

It is possible, as a first approximation, to recognise the main patterns involved in near-surface alteration by tracing the way in which this relatively restricted range of minerals is affected. This is because, firstly, the minerals that have been discussed make up the great bulk of igneous rock, a generally accepted average composition being:

feldspars	60%
amphiboles and pyroxenes	17%
quartz	12%
micas	5%
others	6%

Secondly, it is recognised that all materials making up the lithosphere can be traced back to igneous rock minerals and that therefore the main groups of sedimentary and metamorphic rocks can be considered as special cases arising from this more general one.

In general, the minerals that have been considered formed under conditions of relatively high temperature, high pressure and low oxygen concentration; if H_2O is present it is not in the liquid state. It cannot be expected that such minerals will be stable at the surface of the lithosphere where temperature and pressure are low and oxygen and water are abundant. They will therefore tend to change into minerals which are more in equilibrium with these surface conditions. The general form of the reactions which occur can be determined by applying the 'moderation theorem', which is a more general form of Le Chatelier's principle (Turner and Verhoogen 1960). This theorem states that any reaction will tend to proceed in that direction which will moderate the effect of the initial change. Thus pressure has been lowered; therefore, to moderate this effect, the minerals formed under high pressure will tend to change into minerals which are less dense and hence occupy more space. Temperature has been lowered, so this lowering will be moderated by heat being given out in the process of change from the original minerals to new ones. Due to the low temperature and pressure at which these new minerals form, they will tend to have a very fine grain size and to be of a very mixed nature, as the type formed will depend on very localised conditions.

To this overall process of adjustment to surface conditions the term **epimorphism** can be applied. So far, epimorphism has been discussed only in terms of its starting point and end products, but to understand the formation of soil material it is necessary to consider in detail what happens in intermediate stages. As a point of departure it is possible to suggest certain major processes which it is necessary to investigate:

(a) The primary minerals must break down in some way – this can be termed **weathering**.
(b) The simpler materials resulting from this breakdown must be acted on by surface forces – this can be termed **leaching**.
(c) Reactions between these simple entities will give rise to **new minerals**.

Weathering, leaching and new mineral formation can be regarded as nodes at which many strands meet and they are sequential only insofar as a certain degree of weathering must occur before the other two can start, but in any given mass of material all three will take place simultaneously and react on one another. With this background, each of these processes will be discussed in turn.

2

Weathering

For the purpose of this analysis weathering can be regarded as that part of the epimorphic process in which the minerals of the lithosphere are broken down into simpler entities. This is determined on the one hand by the internal structure of these minerals and on the other by the way in which the environment impinges on them near the lithosphere surface. They will be referred to as internal and external factors respectively and discussed in turn.

Internal factors

Energies of bond formation One of the most important properties in determining the relative resistance of minerals to alteration in the face of changed external conditions is the strength with which an atom or ion is bound to its neighbour. Although little is known at present about the breaking strength of bonds in crystals, energies of formation have been determined for cation–oxygen bonds in silicate glasses (Huggins and Sun 1946) (see Table 2.1) and it is reasonable to assume that bonds requiring the greatest energy to form will be the most resistant to change. On this basis sodium and potassium are very weakly bonded; calcium, magnesium and iron are rather more strongly bonded, although this bonding strength is small compared with that of aluminium, while silicon is almost twice as strongly bound as aluminium. In other words bonds within the silicate framework are very much stronger than for the non-framework cations and it would therefore be very much easier to remove the non-framework cations in the process of weathering.

Bonding energies in standard mineral volumes Given that the bonding within the silicate framework is strong, would there be any difference in the overall bonding strength between the different types of framework? Some approximation to this could be arrived at by taking a standard volume of each framework type in terms of a standard number of oxygen (say 100, as before) and also supposing that bonds of the Si–O type are the only ones that occur in the framework, as was done by Keller (1954). These calculations are shown in Table 2.2 and indicate that there would be an increased resistance to framework breakdown as one proceeds from orthosilicates to tektosilicates. However, such a total breakdown of the framework should not be directly equated with weathering for, in the concept of weathering being discussed here (the production of simpler entities), no such total disruption needs to be considered.

Orthosilicates All that is required to disrupt the structure of minerals such as olivine is the removal of the magnesium and ferrous iron which link the SiO_4 tetrahedra. It is not necessary to disrupt any of the Si–O bonds; this implies that the energy involved in weathering may be considerably less than the energy of bond formation.

Inosilicates Exactly the same argument can be advanced in the case of single chain inosilicates such as pyroxenes for which removal of the bivalent metal cations is sufficient to break down linkage between the chains, while the chains themselves would tend to break down where aluminium is in tetrahedral positions. Thus again, to destroy the inosilicate structure it is unnecessary to break any Si–O bonds. In double chain inosilicates, such as amphiboles, there are a certain

Table 2.1 Energies of formation of oxides of various cations in silicate glasses and minerals (after Keller 1954).

Ion	Kg cal/mole
K^+	299
Na^+	322
H^+ (in OH)	515
Ca^{++}	839
Mg^{++}	912
Fe^{++}	919
Al^{3+} (non-framework)	1793
Al^{3+} (framework)	1878
Si^{4+} (orthosilicates)	3142
(inosilicates, single)	3131
(inosilicates, double)	3127
(phyllosilicates)	3123
(tektosilicates)	3110

number of Si–O bonds involved in the cross-linking between chains; in this case a breakage of some Si–O bonds is required to cause destruction of the amphibole structure, but by no means all such bonds have to be broken. As with orthosilicates, energy values for weathering would be less than the calculated bond energies.

Phyllosilicates   This group represents a special case in the scheme of epimorphism, for phyllosilicates have the same basic sheet structure as most of the secondary clay minerals which form as a result of epimorphism. The bonding energy involved in the weathering of these minerals does not correspond to a great disruption of the original structure, but rather to a modification of it: for instance, the replacement of interlayer potassium by hydroxyls. Therefore, to take the calculated bonding energy of the phyllosilicates as an index of weatherability would be very wide of the mark indeed.

Tektosilicates   In this case there is necessarily a greater breakage of Si–O bonds in order to disrupt the structure, but this would be less for feldspars than for quartz, due to the presence in feldspars of other ions with smaller bonding energies, in particular the tetrahedral aluminium which is absent from quartz.

Thus the figures given for bond strength in a standard volume of silicate minerals, when considered in terms of weatherability, are rather misleading. In general, the correct sequence of mineral weatherability has been arrived at, but the values are too bunched. The values for orthosilicates and inosilicates should be considerably less, while even the values for the tektosilicates will be achieved only if the minerals are reduced to the simple ion form. Due to the special circumstances of their transformation, phyllosilicates should not be considered in the same way; discussion of these transformations will be delayed until the section on new mineral formation in Chapter 4.

Table 2.2   Energies of formation of silicate frameworks

Type	Basic formula	Factor for adjusting to 100 oxygens	Unit bonding energy	Total bonding energy (Kg cal/mole)
Orthosilicate	SiO_4	25	3142	78 550
Inosilicate (single)	SiO_3	33·3	3131	104 366
Inosilicate (double)	SiO_2	36·4	3127	113 823
Phyllosilicate	SiO_2	40·0	3123	124 920
Tektosilicate	SiO_2	50·0	3110	155 500

It can now be said that factors inherent in crystal structures would give rise to a sequence of increasing stability in the face of weathering forces, from left to right as follows:

olivines – pyroxenes – amphiboles – feldspars – quartz

Entities produced by structural breakdown Given this order of mineral stability, attention must now be paid to the entities formed during breakdown. It can be accepted that at the moment of breakdown, all non-framework elements will be in the ionic form; their subsequent behaviour will be best dealt with under the section on leaching (Ch. 3). The question that has to be evaluated now is: 'What kind of fragments result from the breakdown of the silicate framework?'. The fragments produced will be highly unstable due to the number of broken bonds associated with them; thus the period over which such entities exist must be short, thereby making their detection rather difficult. There are three ways in which this problem has been investigated:

(a) experimentally in the laboratory;
(b) theoretically in terms of the energy relations of minerals;
(c) observationally by investigating the weathering products of igneous rocks with modern physical techniques (discussion of this point will be delayed until the section on new mineral formation in Chapter 4).

Laboratory evaluation General framework breakdown was investigated by Murata (1942) by treating silicate minerals with hydrochloric and nitric acids. It was found that either they dissolved completely to yield a clear filtrable solution which set to a firm silicic acid gel, or residual silica remained after the reaction was completed. The minerals that reacted in the first way belonged to two distinct groups: (i) the ortho-silicates and (ii) those tektosilicates with a large amount of tetrahedral aluminium. The second reaction was confined to the inosilicates, phyllosilicates and those tektosilicates with small amounts of tetra-hedral aluminium. These results were interpreted qualitatively as meaning that, if framework units of a small enough size can be produced on initial breakdown, they are capable of forming a colloidal solution and that orthosilicates and aluminium-rich tektosilicates do produce such small particles, whereas inosilicates, phyllosilicates and the aluminium-poor tektosilicates produce much larger fragments. These experiments also pointed out the crucial role of tetrahedral aluminium in this process.

This experimental technique depended on the use of mineral acids which expose minerals to 'weathering' at many times the natural rate. While the results can be used as a general guide to mineral reactions under natural conditions, it would be better to try to get a closer approach to actual weathering conditions. This was done by Frederickson and Cox (1953 and 1954), who studied the effect of pure water at high temperatures and pressures on three tektosilicates: albite, anorthite (both feldspars) and quartz. Apart from the sodium and calcium ions, derived from albite and anorthite respectively, the breakdown products of all three tektosilicates were remarkably similar, consisting of some colloidal-sized materials, and other much larger fragments, which were somewhat altered pieces of the original minerals. This was explained in terms of a mosaic structure for these tektosilicates: there is a strict crystal orientation only within small fragments of a particular mineral and such units are cemented together in a mosaic. The material forming the cement, while being chemically identical with the material inside the mosaic units, would have slightly different physical properties. Being in a strained or disordered condition, the cement would consequently be more soluble than the mosaic units. It was postulated that the cement went into solution as silicate ions which then rapidly polymerised into colloidal form. This caused the mosaic units, whose sizes varied from 0·5 to 5 μ to become detached from the main crystal. In the case of albite and anorthite this made removal of the sodium and calcium much easier, resulting in the expansion of the framework structure that remained behind. This would cause such fragments to be much more reactive than in their original state. Compared with the results of Murata, this more gentle experimentation showed that tektosilicates, no matter how little tetrahedral aluminium they contain, produce a mixture of colloidal-sized materials and much coarser fragments and that insoluble silica is not produced by this means. In view of the fact that this is an approach closer to reality, it can be tentatively concluded that the production of insoluble silica by the breakdown of silicate frameworks would be a very unusual event in nature.

Theoretical evaluation   De Vore (1959) took a theoretical approach to the problem by considering the type of fragment that would be preferentially detached from a perfectly ordered crystal of potash feldspar, assuming that detachment would be easier for those fragments that required fewest bonds to be broken. As a result of detailed analysis he concluded that chain fragments containing both tetrahedral silicon and aluminium would be preferentially formed. An important consequence is that in the formation of clay minerals they are always present together. If silicon and aluminium were in the ionic form, both the very limited solubility of aluminium and the problem of the mutual precipitation of silicon and aluminium to form clay minerals would have to be taken into account. Despite the fact that naturally-occurring feldspars have much greater variability, this work of De Vore at least provides a base from which to appreciate the possibilities.

External factors

Up to this point the discussion of the sequence of mineral breakdown and the entities produced in the process has been based on internal crystal properties. However, no weathering can occur until the environment impinges on these crystals.

The structure of water   In terms of the stability/instability relations of silicate structure, probably the most important difference from the conditions under which the silicate minerals were formed is the presence of H_2O in the liquid state at or near the atmosphere–lithosphere interface. If water were an ideal liquid (that is, one having a close-packed structure), its specific gravity would be about 1·8. It can be concluded that the structure must be far from close-packed. Some indication of the type of structure is suggested in the following line of argument. The latent heat of fusion of ice is relatively low, which implies that the change from ice to water has not been accompanied by any great change in the bonds already present in ice. It has been determined that the oxygens in ice are tetrahedrally co-ordinated. If the change in bonding from ice to water is minimal, it is plausible to postulate that tetrahedral co-ordination persists to a certain extent in water. The tetrahedral co-ordination in ice indicates that a framework structure is present. This is similar to that in the tektosilicates. It has been established (Evans 1964) that the structure of ice is analogous to that of tridymite, the high temperature form of silica. Therefore it can be postulated that water has a pseudo-crystalline arrangement akin to the more compact quartz structure, which would explain the higher density of water relative to ice, and at the same time why a close-packed structure is not attained. It also explains the anomalous increase in density with increasing temperature between 0 and 4°C, for in this range the tridymite-like structure is rapidly breaking down to the more compact quartz-like structure. It is not being suggested that water is a crystalline material, but that at any one instant some water molecules will tend to take up tetrahedral co-ordination.

For a single water molecule, it is known that the size of the oxygen ion is 1·38Å, whereas the distance of the hydrogen ions from the centre of the oxygen is 1·01Å and their size is extremely small; therefore, the hydrogens are effectively buried within the oxygen. Thus the water molecule can be pictured as a sphere which, although neutral, is nevertheless polar due to the asymmetric distribution of the hydrogens, with charges tetrahedrally distributed as shown (Fig. 2.1). This polar character of the water molecule allows **hydrolysis** to take place and this is the most important of the external processes whereby silicates are weathered.

Hydrolysis   Frederickson (1951) explains the process of hydrolysis of feldspars in the following manner. The unsaturated tetrahedra of the feldspar crystal surface give to each crystal face a net negative charge and water molecules in contact with them will orientate themselves so that their positive charges are all towards the crystal surface. This leaves the negative charges of the tetrahedral water molecules facing outwards so that they in turn will attract the positive charges of the next layer of water molecules. However, the perfection of this orientation will decrease fairly rapidly away from the interface. In the few layers where a fair degree of orientation is achieved, the structures of the feldspar and the water are very similar, both having the same basic tetrahedral co-ordination. At the actual interface the water molecule and the

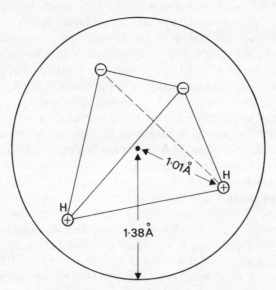

Figure 2.1 The tetrahedral distribution of charge on the water molecule (after Evans 1964, by permission of Cambridge University Press).

oxygen ions of the feldspar are of very similar size; hence there will tend to be a 1 : 1 association on geometric grounds. However, the water molecule will have a double positive charge on account of its orientation, in contrast with the single negative unsatisfied charge on the oxygen of the feldspar. Therefore an excess of positive charges will accumulate along the interface and such an excess will be balanced by a penetration of hydrogen ions into the feldspar lattice. With such a penetration the hydrogen ions will come into regions where there are potassium, sodium or calcium ions, which are all large. For instance, sodium is co-ordinated with about nine neighbouring oxygens. Its unit positive charge is distributed over all these oxygens so that it has a small concentration of charge per unit volume and its bonding with individual oxygens is relatively weak. The hydrogen ion is extremely small and can co-ordinate only two oxygens at the most; the large concentration of charge per unit volume is distributed between only two oxygens and hence would form a much stronger bond with both. Competition between hydrogen ions and the large cations will result in the displacement of potassium, sodium and calcium and their replacement by hydrogen. These displaced cations will move towards the crystal–water interface because of the concentration gradient in this direction; their subsequent movement will be discussed under leaching (Ch. 3). The replacement of these ions by hydrogen involves changes in the feldspar lattice because the hydrogen can co-ordinate with only two oxygens. The remaining seven or eight oxygens previously co-ordinated with the cations have now no such point of attachment and hence they tend to repel one another; this causes an expansion of the crystal at this point which, in turn, increases the strain within the feldspar lattice. This strain is ultimately discharged by the breaking of bonds at the surface of the crystal, with the possible detachment of tetrahedral chains of the type previously discussed.

The effect of hydrolysis on silicate structures The process of hydrolysis has been discussed in detail for feldspars only, but it has general application for all cations that have a relatively weak bonding energy, that is low valency and high co-ordination number Ca^{2+}, Mg^{2+}, Fe^{2+}, Na^+, K^+. These are the first cations to be removed in weathering and the degree of initial weathering of silicate minerals depends on the number and accessibility of these cations. A measure of this is given in Table 1.3, which shows a decrease in the concentration of these cations per unit volume in the order: ortho-, ino-, phyllo-, tektosilicates. Thus, in olivines the great quantity of magnesium and ferrous iron means that they are extremely susceptible to the process of hydrolysis. In these orthosilicates the cations form an essential part of the structure and their removal leads to complete structural breakdown. In pyroxenes and amphiboles, the easy removal of interchain cations leads to the ready detachment of chains and the virtual destruction of their structures. The phyllosilicates such as muscovite contain a univalent 12 co-ordinated potassium in the interlayer position; the cation is very easily removed by hydrolysis. This type of transformation will be discussed in greater detail in the section on new mineral formation in Chapter 4. In tektosilicates such as feldspars, removal of the cations places a greater strain on the bonds remaining, but in quartz, which does not contain any of these easily hydrolysed cations, the process cannot occur.

It can be concluded, therefore, that the process of hydrolysis, insofar as it affects the more weakly bonded metal cations, is responsible for almost complete structural breakdown in orthosilicates and inosilicates, more superficial alteration in phyllosilicates and considerable structural weakening in those tektosilicates such as the feldspars which contain these cations. It would require very much stronger hydrolysis to affect the much more strongly bonded silicon and aluminium cations of the framework structure in a similar fashion, but, if the hydrolysis becomes much stronger, even these cations will be affected and a more complete structural breakdown of the silicate will occur.

Hydrolysis and pH   Hydrolysis is dependent on the concentration of hydrogen ions; therefore, any process which affects the concentration of hydrogen ions will affect the speed of hydrolysis. On average, the pH at the surface of the lithosphere is about 6; that is, slightly acid. Conditions which are liable to increase the acidity (i.e. lower the pH value) are mainly the result of interaction with the biosphere. This will be discussed at greater length when the organic cycle is considered later, but at this point it can be said that breakdown of organic matter leads to the acidification of the environment and hence to an increase in the speed of hydrolysis.

High alkalinity occurs as a result of hydrolysis in the immediate vicinity of many crystal faces when bases such as sodium, potassium, calcium and magnesium are liberated. This has considerable significance in terms of increased solubility of silicon and aluminium and will be considered further in the section on leaching in Chapter 3.

Hydrolysis and temperature   Temperature is of importance, for the higher the temperature the greater the dissociation of water, so that the concentration of hydrogen ions is increased. Hydrolysis will proceed at greater speed the higher the temperature, all other things being equal. However, it must be emphasised that an increase in temperature affects only the rate of reaction, not the type of reaction. As it has been shown (Pickering 1962) that hydrolysis takes place throughout the natural temperature range, there is no reason to postulate a special process of tropical weathering.

Summary

The process of weathering in silicate minerals produces a whole range of materials varying in both size and stability. The sequence ranges from highly stable cations such as potassium and sodium, through colloidal-sized silicate framework fragments, then larger framework fragments, both of which have little stability, to insoluble and unreactive materials such as quartz. The relative speed with which these materials are generated gives rise to a weathering sequence of silicate minerals which, from the most to the least weatherable, is:

olivines – pyroxenes – amphiboles – feldspars – quartz.

3

Leaching

What happens to the varied entities produced by weathering? The degree to which these products are mobile (i.e. subject to leaching) depends upon their ability to form stable ions in aqueous solutions. Whether or not ions will persist in aqueous solution depends on their reaction with the hydroxyl ion of the water molecule (Wickmann 1944). This is determined, in turn, by the ratio between ionic size and valency; in other words, it is a question of balancing geometry and charge, as discussed previously in the case of silicate structures. Cation size determines the number of hydroxyl ions that can be arranged around it, while its valency determines the charge that has to be disposed of.

Ionic size, valency ratio

Large univalent cations such as sodium or potassium can be in contact with six, seven or eight hydroxyl ions. Therefore, any particular hydroxyl ion will be affected by a sixth to an eighth of a unit positive charge, which will not cause any notable distortion in the symmetry of the hydroxyl ion. Bonding is not directional and ions such as these will be stable and persistent in aqueous solution.

The situation is rather different for smaller ions with higher valency. In the case of silicon, only four hydroxyl ions can be arranged around the ion and these four have to take care of four positive unit charges; as silicon is quadrivalent, one whole unit charge is associated with each hydroxyl. This represents a very strong distorting influence on the hydroxyl ion and the result can be pictured in the following way. The binegative charge on the oxygen in the hydroxyl ion may be regarded as being broken into four concentrations of negative charge disposed symmetrically in space; that is, in a tetrahedral form as four half unit negative charges. One of the negative half unit charges is taken up by hydrogen which is pushed out of its position within the oxygen. This leaves the hydrogen with a spare half unit of positive charge. Two other negative half charges are balanced by one of the positive charges of the silicon. The remaining negative half charge attracts the positive half charge of a hydrogen belonging to another hydroxyl group. This produces linked hydroxyl bonds, causing the formation of insoluble hydroxides. Such elements would not persist to any great extent as ions in aqueous solution under natural conditions.

For an element such as phosphorus the situation is different again. This element is quinquevalent, but because of its size only four hydroxyl ions can fit around it. Each hydroxyl now has to cope with $1\frac{1}{4}$ units of positive charge instead of one as in the case of silicon. This excess of cationic charge now competes with the hydrogen in the hydroxyl bonds. This causes the hydrogens to be displaced and, instead of being associated with a particular oxygen, they take up a position equidistant from two oxygens; that is, a hydroxyl bond is replaced by a hydrogen bond. Such bonds are more easily broken and allow the formation of the soluble complex anion phosphate. Consequently it was concluded that, if valency divided by co-ordination number is less than $\frac{1}{2}$, ionic bonds will be formed; if between $\frac{1}{2}$ and 1, hydroxyl bonds; and greater than 1, hydrogen bonds (Wickmann 1944).

Ion mobility groups

It can be inferred from Figure 3.1 that, when subjected to leaching, elements belonging to groups I and III will be relatively much more mobile than the elements of group II. This explains one of the basic characteristics of epimorphism: the relative enrichment of the residual masses in silicon, aluminium and ferric iron. Thus the major elements present in a weathering mass are divisible into two groups: (a) those which occur as simple ions and are highly mobile, such as calcium, sodium, magnesium, potassium and ferrous iron; and (b) those which tend to form insoluble hydroxides and are relatively immobile, such as silicon, aluminium and ferric iron. The complex ions of group III, apart from carbonates, are not major components of the system.

The highly mobile cations are the same as those which are readily freed from silicate minerals; their high mobility and ready release from silicates are both attributable to their large co-ordination numbers and low charges. It is difficult to make a general statement about the relative mobility of ions within this group; in particular circumstances any one of them could be classified as being the most mobile. Certain of these differences are caused by feedback reactions with other products of epimorphism: for example, potassium is absorbed to a much greater extent by clay minerals than is sodium.

Silicon, aluminium and iron

Rather more detailed discussion is necessary for these major immobile elements. In the case of iron, the ferrous form belongs to the highly mobile elements of group I, while ferric iron belongs to the immobile elements of group II (Fig. 3.1). Whether one or the other occurs under natural conditions is dependent upon the activity of protons and electrons, which are measured in terms of pH and Eh respectively. An example from Hem and Cropper (1959) is given in which the stability field for the aqueous ferric–ferrous system is given in terms of these two components (Fig. 3.2). Within this Figure the general pH–Eh range of the natural environment has been shaded, showing that for two-thirds of this area insoluble ferric hydroxide would be formed, which means that ferric iron in excess of 0·01 ppm can only occur in suspension. Only in the remaining one-third of the area would soluble ferrous ions occur and at a maximum this would not exceed 100 ppm (Eh 0·3, pH 5·0). Thus under natural conditions of leaching, practically all the iron is

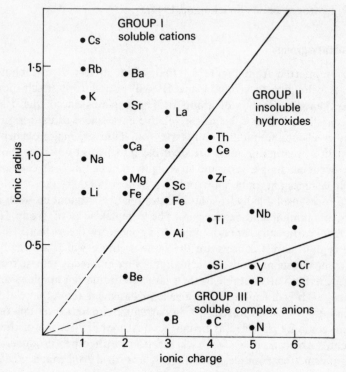

Figure 3.1 Ionic potential of elements.

quickly changed to the ferric state within a very short distance of being freed from primary minerals by weathering, becoming an inert residual product which reacts further only under special conditions.

Silicon is more complex, as its solubility rises considerably with increasing pH. Figure 3.3 shows the increase in solubility of amorphous silica at a temperature of 22°C as pH increases from 5 to 11, as determined by Okamoto *et al.* (1957). From this it can be seen that from pH 6 to 9 solubility remains relatively steady at about 200 mg/l but between pH 9 and 10 it rises very quickly to about five times this value. Taking 10°C and pH 6 as average values at the surface of the earth, 125 mg/l of silica would be the expected content in natural waters, since there is no lack of silica at the surface of the earth. However, the average content of silica in stream water has been found to be 13 mg/l (Davis 1964) and it occurs in the monomolecular form (H_4SiO_4): that is, there are no colloidal-sized entities. This disparity between the actual and expected amounts of silica has been explained as being due to

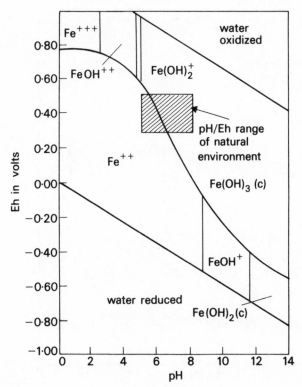

Figure 3.2 Stability-field diagram for aqueous ferric–ferrous system (after Hem and Cropper 1959).

aluminium, which is also freed from crystal surfaces at the same time as silicon but in much smaller amounts. These smaller amounts are, however, sufficient to cause the precipitation of most of the silicon once it has moved away from the zone of very high pH at the crystal surface.

Thus the sequence suggested for both silicon and aluminium is as follows. At the crystal surface under the highly alkaline conditions resulting from hydrolysis, a certain amount of silicon and aluminium go into solution; at the same time this causes the breakdown of the original silicate structure and the production of framework fragments of various sizes, most probably of a chain form (for reasons discussed earlier). As this silicon- and aluminium-rich solution moves away from the crystal surface, lower pH conditions result in the solubility being exceeded and the silicic acid polymerises into colloidal-sized entities. These are almost completely precipitated by the small quantity of aluminium present, leaving in solution only a small amount of the

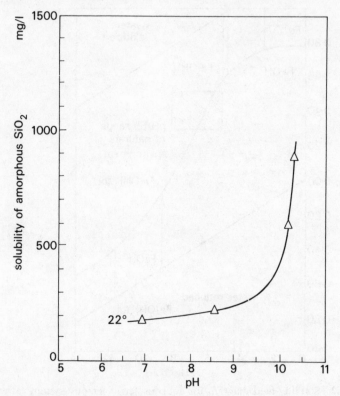

Figure 3.3   Solubility of amorphous silica (derived from Okamoto *et al.* 1957).

monomolecular form to be leached out of the system. This has considerable implications for both leaching and new mineral formation for, no matter whether the framework fragments go into solution or remain as chains, the silicon and aluminium remain intimately associated with one another; it would be better to consider them in this way rather than as independent entities.

Relative mobility scale

From these considerations it is possible to arrange the major cations in a sequence of relative mobilities:

1. Calcium, sodium, magnesium and potassium – occur as simple, highly mobile ions, readily removed in the process of leaching.
2. Silicon and aluminium – immobile, moving over only short distances as colloidal entities; a small amount of silicon, as H_4SiO_4, can move in solution over great distances.
3. Iron – in the highly oxygenated surface areas occurs in the ferric form, which is highly immobile and inert.

4

New mineral formation

The nature of secondary phyllosilicates

The new minerals produced at the surface of the lithosphere in response
to changed conditions have, in most cases, a phyllosilicate or sheet
structure. Weathering and leaching can be regarded as processes
whereby cations are removed relative to oxygen and it would appear
that under these conditions the only silicate structure that can attain
stability is the phyllosilicate. It is possible to differentiate two main
lines by which such phyllosilicates can be produced. The first is from
primary micas, such as biotite and muscovite, which already have a
phyllosilicate structure. In this case the change does not necessarily
involve a fundamental structural alteration and the main problem is the
way in which one phyllosilicate turns into another. The second line is
from other silicates that do not have a phyllosilicate structure and this
necessarily involves a basic rearrangement of structure.

Before going further with this discussion it is necessary to say something more about phyllosilicate structure, particularly regarding that of
clay minerals. The phyllosilicate structure was discussed in Chapter 1
and illustrated in Figure 1.9, where it was shown that the basic structural unit consists of two sheets of silicon tetrahedra with the apical
oxygens pointing towards one another. These two sheets are joined
together by either divalent or trivalent ions such as magnesium and
aluminium, bonded via oxygens and hydroxyls. The external sheets with
the silicon and a certain amount of aluminium in tetrahedral co-
ordination are referred to as **tetrahedral layers**. The single internal sheet
is referred to as the **octahedral layer,** as the divalent and trivalent ions
are co-ordinated with six oxygens and hydroxyls spatially distributed
in the form of an octahedron. This normal phyllosilicate with one
octahedral and two tetrahedral layers is referred to as a 2 : 1 mineral.
The 2 : 1 structure also occurs in many clay minerals such as montmorillonite, vermiculite and illite. However, in the kaolin group of clay

minerals, the basic structure consists of one tetrahedral and one octa-hedral layer only. That is, they have a 1 : 1 structure.

The alteration of primary micas

There have been numerous investigations in both the field and the laboratory of the way in which primary micas alter under epimorphic processes. Unfortunately, for the most part these investigations have produced a mass of contradictory results not only regarding the relative resistance of the original micas but also with respect to the alteration products. This is due to a combination of factors. Compared with the alterations undergone by other silicate minerals during epi-morphism, the changes undergone by micas generally are compara-tively subtle; to a certain, but largely unknown, extent these slight changes are controlled by the chemical composition and structure of the micas and, as mentioned in Chapter 1, the possibilities of ionic substitution within the micas are considerable. Therefore, in order to rationalise this complex and confused – but highly important – area of epimorphism, it is necessary to restrict consideration to micas of known structure and chemical composition being operated on by controlled environmental factors. The first step in this rationalisation is to distinguish two main types of mica. The first comprises those micas such as muscovite that contain trivalent ions, particularly alu-minium, in the octahedral layers. In this case the charge on the crystal is balanced by only two out of every three octahedral positions being occupied – therefore such micas are called **dioctahedral.** The other type comprises the **trioctahedral** micas such as biotite, that have all octa-hedral positions filled by divalent ions, e.g. magnesium and ferrous iron. It is proposed to examine examples of muscovite and biotite alteration in which the chemical composition of the original materials is kept uniform and environmental conditions are controlled. By taking such an approach it will then be possible to appreciate the more complex situations in which both sets of factors are variable.

The alteration of muscovite The first example is from New Zealand where a muscovite mica of relatively uniform composition is a con-stituent of the greywackes belt that runs from north to south for a considerable length of both islands. By a careful selection of sites it was possible to study the general trends of the alteration of this mica under a range of environmental pressures (Fieldes and Taylor 1961). The main stages in this process are an initial change to illite and then, depending on whether leaching is strong or weak, through clay vermiculite to metahalloysite or to montmorillonite (Fig. 4.1). From Table 4.1 it can be seen that the most obvious change is the removal of interlayer potassium by hydrolysis, for in terms of potassium atoms per unit cell it is 1·5 in muscovite, 1·0 for illites, 0·5 for clay vermiculite and is absent in montmorillonite and metahalloysite. The potassium is responsible for the cohesion between the individual mica sheets, so its removal causes a weakening in this bonding, as is shown in the column of X-ray basal spacings which gives the unit distance over which the basic phyllosilicate pattern is repeated (Fourth column, Table 4.1). In the primary micas this repeat distance is 9·8Å which increases to 17·8Å in montmorillonite. Such weakening of bonds allows exfoliation to occur, producing thinner and thinner phyllosilicate plates. As well as decreasing in thickness, the diameter of the fragments decreases very rapidly in montmorillonite from 1 μm to between 0·1 and 0·5 μm. Such processes produce more surface area per unit volume of material. These surfaces tend to be highly reactive, as is shown by the increase in cation exchange capacity (the sum total of exchangeable cations that the clay mineral can absorb, expressed in milliequivalents per hundred grammes of material) from very low in primary micas to 100 meq for clay vermiculite and montmorillonite. The changes involved in the line from muscovite through illite to montmorillonite and those in the other line of descent as far as clay vermiculite are essentially superficial. The basic phyllosilicate structure has remained untouched; this is reflected in the stable SiO_2/Al_2O_3 ratio of 3 ± 1, for silicon and aluminium are the two elements involved in the framework structure. In other words, alteration has been confined to the removal of non-framework cations resulting in the comminution, but not the destruction, of the phyllosilicate structure.

However, under conditions of very strong leaching, the clay vermiculite structure is subject to a more fundamental breakdown to metahalloysite, a disordered form of kaolinite. A 2 : 1 phyllosilicate is changed into a 1 : 1 phyllosilicate. Such a major change in the framework is indicated in Table 4.1 by a change in the SiO_2/Al_2O_3 ratio to $1\frac{1}{2} \pm \frac{1}{2}$ for metahalloysite as against 3 ± 1 for the 2 : 1 phyllosilicates.

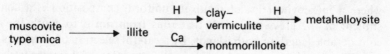

(H) indicates strong leaching at low pH
(Ca) indicates restricted leaching at relatively high base saturation

Figure 4.1 Course of muscovite alteration (after Fieldes and Taylor 1971)

Table 4.1 The alteration of muscovite (after Fieldes and Taylor 1961).

Mineral	Potassium atoms per unit cell (approx.)	SiO_2/Al_2O_3 ratio of mineral	Cation exchange capacity (meq per cent)	X-ray basal spacings (glycerol)	Particale size (μ)
Primary micas	1·5	3±1	very low	9·8	1
Illite	1·0	3±1	30	10·2	1
Clay vermiculite	0·5	3±1	100	14·0	1
Monmorillonite	0	3±1	100	17·8	0·1–0·5
Metahalloysite ($+SiO_2$)	0	$1\frac{1}{2}\pm\frac{1}{2}$	very low	7·4	0·2–1·0

To achieve this change the original mica structure of two tetrahedral sheets enclosing one octahedral sheet, still in this form in clay vermiculite, has to be changed into a kaolinite structure of one tetrahedral and one octahedral sheet. Additionally, in the mica structure a certain amount of aluminium occurs in the tetrahedral positions while in the kaolinite structure of metahalloysite, silicon occurs only in these sites. Therefore, to change from one to the other requires a tetrahedral layer to be stripped off and the aluminium in the remaining tetrahedral layer to be removed. A possible mechanism suggested by Coleman (1962) is shown in Figure 4.2. At the bottom left-hand corner of the diagram is shown a clay particle with a mica structure. Under strong leaching conditions exchangeable bases are removed, leaving hydrogen as the dominant exchangeable cation. This causes great strain within the mica lattice which is relieved by the replacement of the hydrogen by aluminium from within the lattice, producing structural breakdown. The occurrence of exchangeable aluminium means that a certain amount of aluminium must go into solution. Due to the size/charge relationship of

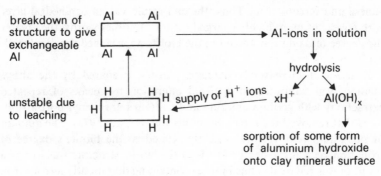

Figure 4.2 Removal of aluminium from octahedral sites (after Coleman 1962).

aluminium ions they would be almost immediately precipitated as aluminium hydroxide in octahedral co-ordination which would combine with one or other of the tetrahedral layers to form a disordered kaolinite, namely metahalloysite. The metahalloysite in this case is in the form of flat sheets, because it is interlayered with residual mica structures having a flat sheet form which acts as a template for metahalloysite development.

The alteration of biotite   The problems involved in the study of biotite are greater than in the case of muscovite, for the range of possible ionic substitutions in the octahedral layer is greater and the variety of possible alteration products is considerably increased. In particular, there has been divergence of opinion regarding the relative resistance of biotite to alteration. It was realised that this last problem was connected with the oxidation state of the iron, for biotites rich in ferric iron seemed to show more resistance to the processes of alteration. The difficulty was that in laboratory experiments it has been found impossible to produce the oxybiotites that occurred in the field without altering the biotite structure to that of vermiculite or montmorillonite, so making it impossible to differentiate between oxidation and alteration.

The controlled laboratory oxidation of biotite without structural alteration was accomplished by Gilkes *et al.* (1972) and this finally allowed an evaluation to be made of the effect of the oxidation state of the biotite on its subsequent alteration. A biotite was oxidised by shaking one of standard grain size with saturated bromine water at an elevated temperature in a sealed glass ampoule; the longer the period of shaking, the greater was the degree of oxidation. In the process the octahedral iron was oxidised to the ferric state but, as the biotite structure had not altered, charge balance could only be maintained by the expulsion of ions from octahedral sites such as magnesium, manganese and ferrous iron. Thus, the end result was an octahedral layer with fewer but more highly charged cations. In other words, the greater the degree of oxidation, the more the biotite approached the dioctahedral form.

A series of increasingly oxidised biotites, prepared by the above method, was subjected to artificial alteration by means of repeated extractions with sodium chloride and sodium tetraphenyl boron solutions that removed the interlayer potassium (Gilkes 1973). The amount of potassium removed was directly related to the biotite's degree of oxidation, as was the degree to which the biotite structure was lost as a result of this removal. Thus biotites containing dominantly ferrous iron released a great deal of potassium and in the process were largely changed into vermiculite, while highly oxidised biotites lost relatively little potassium and retained their biotite structure to a much greater extent. The same sequence of potassium extraction and structural alteration related to oxidation state was observed under much more strongly acid conditions when hydrochloric acid was used as an extraction agent instead of sodium chloride and sodium tetraphenyl boron (Gilkes *et al*. 1973).

As a result of this work it is now possible to explain the variable reactions of biotite to environmental conditions for, where a ferrous iron-rich type trioctahedral biotite is involved, it approaches the pyroxenes and amphiboles in its ease of weathering, but where a ferric iron-rich oxybiotite is involved it approaches a dioctahedral mica, such as muscovite, in its resistance to alteration; in fact, the reactions of biotite demonstrate that it is both a ferromagnesian mineral and a phyllosilicate.

The formation of secondary phyllosilicates from other minerals

The derivation of secondary phyllosilicates from other types of silicate structure is now considered. This is best discussed from two angles:

(a) reactions involving the breakdown of a perfectly ordered tekto-silicate;
(b) Reactions involving materials having extreme internal disorder.

Most naturally occurring materials are somewhere in the middle but, unless the process is discussed in terms of these extremes, it is difficult to understand the more normal intermediate situation.

Perfectly ordered tektosilicates   Given a perfectly ordered mineral, the theoretical approach initiated on the section on weathering in Chapter 2 (De Vore 1959) can be continued. As stated previously, in a potash feldspar the unit most likely to be detached from the commonly developed crystal faces is of a chain type. It was then shown that sheet structure could be developed from basic chain units with relatively little rearrangement. The stability of such a structure is dependent upon the availability of cations for the octahedral layer; this is controlled by the leaching conditions of the environment. The high cation concentrations associated with weak leaching conditions tend to give rise to minerals such as montmorillonites. Under the low cation concentrations associated with strong leaching, aluminium, being relatively immobile,

New mineral formation

is frequently the only cation available for octahedral co-ordination which leads to the formation of a kaolin type mineral. In the initiation of this sheet structure the surface of the crystal undergoing attack could very well act as a template, for the surface layer consists of 'mica-like' tetrahedra as a result of broken bonds. Any cations that can be octahedrally co-ordinated and which are freed by the process of weathering can become bonded to such a surface to which in turn the chain fragments can become attached. Once such a nucleation centre has become established, phyllosilicate formation will continue, even though the initially bonded cation may not retain its attachment to the feldspar surface for very long.

Highly disordered volcanic glass In the case of materials with very disordered silicate structures the surface occurrence of silicon and aluminium would not be at all uniform and periodic. Therefore, fragments detached from the surface of such materials would be of a heterogeneous nature and there would not be enough ordered chain units capable of forming sheet structures directly. Even the few chain units that would form under these conditions would be prevented from polymerising by the presence of other groups not used in this process; being unstable, the chain units would tend to break down further into much smaller units before they could be stabilised into a growing mineral. In these circumstances minerals of an extremely disordered type would be expected to develop. Again, New Zealand provides a suitable example where extremely disordered silicates have been subjected to varying degrees of epimorphic pressure. In the North Island there is a series of rhyolitic and andesitic ash showers varying in age from Upper Miocene to Holocene. These showers contain volcanic glass and feldspars, both of which have highly disordered structures from being quenched during their explosive transfer from magma to ash shower. The type of clay mineral developed on each of these ash showers as a result of epimorphic processes can thus be related to the time after the ash shower was deposited. Using X-ray diffraction, differential thermal analysis, infra-red absorption and electron micrography, Fieldes (1955) has investigated the clays developed in seven soils derived from these ash showers. X-ray diffraction showed the clays to be largely amorphous, particularly those from the younger ash showers, and these materials were distinguished from one another by the other three methods. It was concluded that a sequence of new minerals developed, showing an increase in crystal size and degree of organisation with time (Fig. 4.3).

The overall trend is from allophane, a highly disordered 1 : 1 phyllosilicate, to metahalloysite, a much less disordered 1 : 1 phyllosilicate and possibly, after a much greater period of time, to an ordered kaolinite. This sequence can be interpreted as follows. Under the prevalent conditions of strong weathering, any alkali and alkaline-earth cations would be quickly removed by hydrolysis and leaching. The disordered framework structure of the volcanic glass and feldspars would break down to fragments of very small size, possibly single tetrahedra. Because of this and high alkalinity at the mineral surface due to hydrolysis, a considerable quantity of both silicon and aluminium would go into solution as ions. The very strong fall in pH away from the site of hydrolysis leads to the co-precipitation of both silicon and aluminium within a very short distance. Given this speed of reaction over a very short distance and the small size of the initial fragments, it is to be expected that the degree of disorder in the product would be great. This has been proved to be so for allophane; under moist conditions the allophanes may be visualised as gel-like fragments of aluminosilicates held together by random cross-linking at a relatively small number of sites. The resulting structure is very open and porous and consequently retains a great deal of water. As long as the water is not removed, shrinkage cannot occur and further cross-linking is prevented. Thus, under continuously moist conditions, allophane will

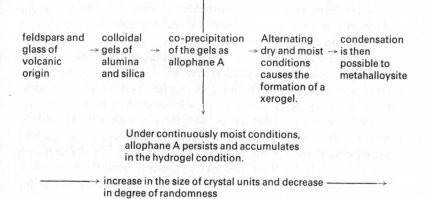

Figure 4.3 New mineral formation from primary minerals with disordered structure (derived from Fieldes 1955).

persist and accumulate. However, if water is lost as a result of alternating wet and dry periods, shrinkage and increased cross-linking occur, leading to a greater order and a change towards metahalloysite. In contrast to the metahalloysite derived from micas, which is largely in the form of flat sheets, that derived from allophane occurs mainly as tubular rolled up sheets because there are no mica fragments to act as templates.

Working from this special situation where allophanes are dominant, Fieldes (1966) was able to show that allophane also develops at an early stage in the epimorphism of basalts following the breakdown of olivines, amphiboles and pyroxenes to fragments similar in size to those produced by volcanic glass and disordered feldspars.

Perhaps the most significant point about the three modes of development is that, in each case, under leaching conditions a mineral of the kaolinite group is produced – generally metahalloysite. The reason is that aluminium is the only element both available and able to go into octahedral co-ordination. This is reflected in the almost ubiquitous presence of kaolinites in soil clays.

The role of non-framework ions

Up to this point discussion has been restricted to the formation of the aluminosilicate sheet structures. Even though this represents the major problem of new mineral formation, there is also the associated problem of how all the other elements behave. This is best approached by considering how each of the three major groupings of elements considered in the process of leaching (Fig. 3.1) reacts. The elements of group I which can form simple cations are held in an exchangeable form on the surfaces of clay minerals. The number is decided mainly by the amount of substitution in the octahedral layer of the clay mineral. Thus montmorillonites hold the greatest quantity, illites intermediate amounts and kaolins the least. In general, the most common exchangeable cations are calcium, magnesium, potassium and sodium, divalent cations generally being much more abundant than monovalent. As discussed previously, whether or not these simple cations will be included within the clay mineral lattice depends upon the leaching environment. If there is an abundance of such ions, the divalent ions occur in the octahedral layers of 2 : 1 clay minerals while the monovalent ions, such as potassium, will be confined to interlayer positions binding the clay platelets together.

The elements of group II, which form insoluble hydroxides, behave in the same way as silicon and aluminium; hence they become occluded within the clay minerals as the growth of the lattice proceeds. As can be seen from Figure 3.1, this includes a wide range of elements.

The elements of group III are capable of forming complex anions such as phosphates, molybdates and vanadates. For these to become associated with clay minerals, the development of positive charges at the clay surface is required. How this occurs is not too clear but it appears to be connected, first, with the degree of disorder in a particular clay mineral (the greater the disorder, the greater the anion retention; thus allophane is capable of more anion retention than kaolinite) and, secondly, with the occurrence of aluminium and iron in tetrahedral co-ordination. For aluminium this happens in montmorillonites and for iron in certain hydroxides, although the factor of disorder may also be involved, particularly where the material has been freshly precipitated.

An integration of weathering, leaching and new mineral formation

The processes of weathering, leaching and new mineral formation can be integrated in the manner shown in Figure 4.4 (Crompton 1960 and 1962). In this diagram increasing intensity of leaching is plotted against increasing intensity of weathering. In very general terms, the type of clay mineral produced is dependent upon the balance achieved between weathering and leaching. An environment rich in cations and alumino-silicate fragments will give rise to 2 : 1 lattice clays. Where the intensity of weathering and leaching has increased to a level at which few mobile cations are present, the alumino-silicate fragments will form a 1 : 1 lattice clay. In the position of extreme weathering and leaching, no

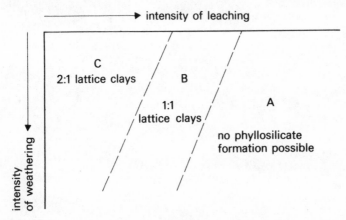

Figure 4.4   Weathering/leaching as an epimorphic index (derived from Crompton 1960).

cations and no alumino-silicate fragments will remain; therefore there is no possibility of any secondary phyllosilicate development. All that will remain is the residual product of weathering (quartz) and those of leaching (iron and aluminium oxides).

The overall result of these three processes is material of very contrasted particle size. The quartz residue from weathering is generally greater than minimum sand size (0·02 mm), while the newly formed materials, whether of a crystalline or amorphous nature, are all finer than maximum clay size (0·002 mm). Therefore, the more that any material has been subject to the processes of weathering, leaching and new mineral formation (the amount being determined by the product of time and the intensity of operation), the greater will be the tendency towards a bimodal distribution of particle size, with a decreasing proportion of particles in the silt-size grade (0·02–0·002 mm).

5

Soil fabric

Definition of fabric

So far discussion has been restricted to the way in which mineral particles are produced from primary igneous rock minerals as a result of the processes of weathering, leaching and new mineral formation. In combining to form soil material these mineral particles, as well as having a particular composition and size, have a spatial relationship with one another, i.e. a **fabric** is developed. In this case, the term 'fabric' is being used in a general sense to include concepts such as structure and pedality.

Residual and newly formed particles   There are two types of mineral particle involved:

(a) residuals, such as quartz, which are inert and coarse-grained;
(b) newly formed minerals, such as the hydrous oxides and clay minerals, that are of much smaller grain size and have a considerable degree of structural disorder. (This means that such materials are more easily moved mechanically and are chemically more reactive than the residual minerals.)

Gap between laboratory and field studies   When an attempt is made to explain the formation of soil fabric from the ingredients provided by these processes, the enormous gap between laboratory work dealing with single-cation, single-electrolyte, single-mineral systems and the complexities of natural soil material becomes very apparent. This proviso must be kept in mind throughout the following discussion in which an attempt is made to bridge this gap.

Laboratory studies

Flocculation and deflocculation   In dilute suspensions of a clay mineral in water the character of the suspension is determined by the type of

clay mineral, the size of the particles, the kind and quantity of exchangeable ions occurring on the surface of the clay mineral and the
kind and concentration of ions in the water. It has been observed that
if a fine-grained clay mineral with sodium as its exchangeable cation
is shaken up in distilled water, it will usually disperse into individual
particles. In this state the clay mineral is said to be **deflocculated.** If, to
such a suspension, sufficient electrolyte is added, the individual particles
will begin to stick to one another, giving rise to a loose aggregate or
floc. The clay mineral is then said to be **flocculated**. This change in
behaviour of the clay particles is explained in terms of a shift in the
balance between forces of repulsion and attraction consequent upon
addition of the electrolyte.

Diffuse double layer   Clay minerals carry a net negative charge on their
basal surfaces due to isomorphous substitutions within their lattices,
mainly in the octahedral layer. This negative charge is balanced by the
absorption of cations on the basal surfaces. When a clay particle is
placed in distilled water, there will be a tendency for these cations to
move away from the particle surface into the water, in order to reduce
the concentration gradient between the two. The majority, however, do
not move very far away from the particle surface and there is a diffuse
clustering of ions about the clay particles. The clay surface with its
negative charge is regarded as one layer, while the surrounding cluster of
cations is regarded as another; together they are referred to as a **diffuse
double layer.** Particles in this state repel one another because the outer
parts of the double layers have the same net electrical charge. The range
and effectiveness of the repulsive force is controlled by the thickness of
the double layer which is thought to vary from 50Å to 300Å. The force
of repulsion is at a maximum at the particle surface and decreases
gradually away from it, as shown in the top part of Figure 5.1. The
attractive force between particles is very great at their surfaces – much
greater than the repulsive force – but the rate of decrease of this force
away from the surface is also very much greater than that of the
repulsive force, as shown in the bottom part of Figure 5.1. The resultant
of the two types of force is also shown. When the inter-particle distance
is reduced to something less than 20Å the attractive force between
particles is dominant and hence flocculation will occur, whereas at
distances in excess of about 20Å the force of repulsion will be dominant
and deflocculation will occur.

Electrolyte effect   Small inter-particle distance will be achieved by any
factor which reduces the thickness of the diffuse double layer. Floc-

culation was seen to result from the addition of an electrolyte to a deflocculated sodium clay for this would increase the concentration of cations in the water and hence decrease the concentration gradient between the surface of the clay particle and the surrounding liquid. This, in turn, would reduce the tendency for exchangeable cations to diffuse away from the clay mineral surface and, in effect, would reduce the thickness of the double layer to such an extent that the inter particle distance would be reduced to the point where the attractive force becomes dominant.

Cation effect The thickness of the diffuse double layer is also dependent on the type of cation occurring in the exchangeable position. Sodium and potassium are responsible for diffuse double layers of the greatest thickness, while those formed by calcium and magnesium are less thick and by aluminium very much less again. This, of course, is related to the valency of the cations concerned and also appears to be related to the ease with which cations will dissociate from the clay surface. Sodium seems to dissociate to the greatest extent and will thus form the thickest double layer, while calcium is much less active. As a corollary to this it was found that, at equivalent concentrations of electrolyte, divalent cations decrease the thickness of the double layer more strongly than monovalent, and trivalent more strongly than divalent.

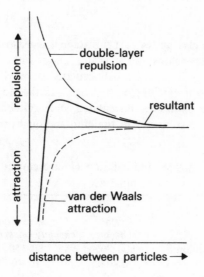

Figure 5.1 Forces between basal surfaces of adjacent clay mineral particles.

Explanation of flocculation/deflocculation From this it can be concluded that, with fine-grained clay mineral particles, exchangeable
monovalent cations will tend to induce a deflocculated state in dilute
suspensions and that this tendency decreases as the valency of the
cation involved increases. However, this basic idea involves certain
assumptions: namely, that the basal planes of all particles are parallel
to one another and that the distance between particles is uniform.
Experimental verification of the double layer theory seems to be restricted to the clay mineral montmorillonite (which tends to be very
fine-grained) in water, with sodium chloride as the electrolyte. In the
case of other clay minerals and other electrolytes, the degree of deviation
from the result expected from the double layer theory becomes quite
considerable. That this is to be expected is shown in Table 5.1, which
gives the specific surface and the cation exchange capacity of montmorillonite, illite and kaolinite. The decrease in specific surface of up
to 150 times from montmorillonite to kaolinite, paralleled by a similar
decline in cation exchange capacity, means that the negative charge on
the basal surfaces becomes relatively much less important in controlling
the behaviour of the clay particle in the same direction. Consequently,
the charges associated with broken edges of clay particles become
relatively more important for kaolinite. Some of these charges are
positive, which causes the positively-charged edges to be attracted to the
negatively-charged basal surfaces, thereby leading to an edge-to-face
arrangement of particles. Such a reaction would tend to create a
flocculated state.

Application of thick clay pastes At this stage it is reasonable to suggest
that, in dilute suspensions, 2 : 1 lattice clays with sodium as the
dominant cation would assume a deflocculated state with ease, while a
1 : 1 lattice clay with calcium or hydrogen as the dominant cations
would more easily assume a flocculated state. Other factors which
would influence the result, such as the kind and quantity of anions, the

Table 5.1 Specific surface and cation exchange capacity of some clay
minerals.

	Specific surface m^2/g	Cation exchange capacity ($meq/100\ g$)
Montmorillonite	600–800	80–150
Illite	65–100	10–40
Kaolinite	5–30	3–15

pH and organic matter, have not been evaluated so far. The type of fabric produced in laboratory experiments on flocculation and deflocculation has, for the most part, been inferred indirectly from observations of such things as sedimentation volume. In very few cases has the fabric been measured directly by X-ray, optical or other means. It is generally thought, however, that the flocculated state is associated with an edge-to-face arrangement of particles, while the deflocculated state is due to a parallel arrangement of the particles, either within limited domains or throughout considerable bodies of material. Accepting these ideas regarding basic structures, it is possible to understand the behaviour of thick clay pastes which are closer to actual soil materials than the dilute clay suspensions (Russell 1961). Having a structure with a preferred orientation, a deflocculated clay paste flows more easily than a flocculated one. On drying, the deflocculated paste forms a hard, dense mass of material which, on re-wetting, quickly disperses into a deflocculated suspension. However, the result of drying a flocculated paste is a porous, cracked mass of material. Re-wetting this causes the individual aggregates produced by the drying to swell, but to retain their coherence; they do not redisperse. Thus, water-stable entities are produced by flocculated material and water-unstable entities by deflocculated material. Therefore, it can be concluded that soil material tending towards the deflocculated state will be more prone to movement, reorganisation and concentration, whereas soil material tending towards the flocculated state will tend to behave in a much more stable manner. Furthermore, the first type of material, with its preferred orientation, will tend to be dense, while the second, with its lack of preferred orientation, will tend to be much more porous.

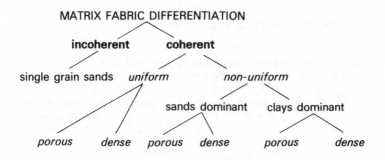

Figure 5.2 Matrix fabric differentiation.

Field application

A method of soil fabric evaluation in the field will now be given. This will be followed by a section in which particular fabric types are related to the processes discussed in this and previous chapters. However, it must be kept in mind that the complexities involved in soil fabric development in the field are enormous compared with those involved in the laboratory studies discussed in the first part of this chapter and, therefore, any suggested interrelationships cannot be regarded as more than tentative.

Soil fabric description in the field   In observing soil material in its natural – that is, undisturbed – state, account has to be taken of the way in which the solid particles are arranged with respect to one another and also the arrangement of the complementary spaces. This can be referred to as the **matrix fabric**. In addition, within particular zones planes of weakness occur which are expressed in the form of planar voids on a macro-scale.

In the differentiation of types of matrix fabric the main criterion used is that of coherence among soil particles (Fig. 5.2), which differentiates between single-grain sands on the one hand and coherent soil material on the other. Within coherent soil material further differentiation is possible, depending on whether the matrix fabric is uniform or not and also on whether it is porous or dense.

With regard to planar voids (Fig. 5.3), there is an essential difference between those soil materials in which planar voids are absent or, at most, sporadic, and those in which the voids form an integrated pattern, so that three-dimensional entities, or peds, can be detached. This is the basis of the main differentiating criterion. Within those materials that

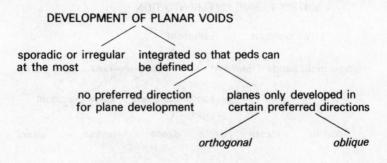

Figure 5.3 Planar void differentiation.

have an integrated system of planar voids it is possible to differentiate between those in which the voids are developed in certain preferred directions and those in which no such pattern can be detected. In the first group different types of pattern, such as orthogonal or oblique, can be recognised and in some cases it is possible to relate such patterns to bedrock joints and bedding planes.

Depending on the ease with which peds can be detached from the mass of the soil, it is possible for each integrated planar void pattern to be weakly, moderately or strongly developed. Those soil materials with no preferred orientation of planar void development generally produce polyhedral peds, while orthogonal void patterns produce blocky peds. Oblique patterns can produce a range in both size and shape of ped, depending on their complexity.

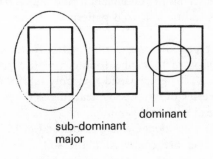

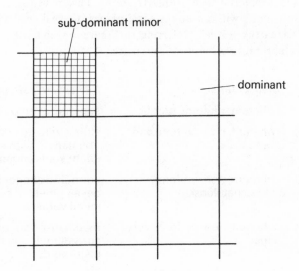

Figure 5.4 Subdominant major and minor void patterns.

The planar void pattern that is immediately obvious to the observer is the dominant one, but in some cases it can be seen that the dominant peds can be grouped into larger entities defined by more weakly-developed planar voids. Thus, polyhedral peds may be associated together into larger, but more weakly-developed, prismatic peds, or may only be larger and rather weaker development of the same type of ped that is dominant. Such a void pattern can be thought of as being a **subdominant major pattern** (Fig. 5.4).

More commonly, a dominant ped is itself cut by a more weakly-developed void pattern which defines smaller peds. In this case the void pattern can be regarded as a **subdominant minor pattern**.

This method of approach means that emphasis is transferred away from pedality as such, which presents pedologists with so much difficulty, to the much more fundamental and more readily analysable factors of matrix fabric and void pattern.

Common fabric types   By considering matrix fabric and void pattern together it is possible to define four common types of soil fabric (Fig. 5.5). It should be emphasised that only fabrics produced by the epimorphic processes of weathering, leaching and new mineral formation are being dealt with at this point.

Fabric type (a) is developed where oxides and/or hydroxides in a fine state of subdivision are common in the soil. In this condition the finely divided grains are capable of forming bridges between any residual material such as quartz grains. This gives rise to an extremely porous mass without any natural planes of weakness and is referred to as 'earthy fabric'. Referring to Figure 4.4, materials necessary for this fabric would be produced in area A.

	Matrix fabric	Planar void pattern
(a)	coherent, uniform, porous	sporadic or irregular at most
(b)	coherent, non-uniform and porous	strong with no preferred orientation, with sub-dominant minor voids common
(c)	coherent, non-uniform, somewhat dense	strongly orthogonal slight development of sub-dominant minor voids
(d)	coherent, non-uniform, very dense	moderate orthogonal, considerable sub-dominant major voids

Figure 5.5 Common fabric types.

Fabric type (b) is developed from soil material in which the 1 : 1 type clay mineral lattice is dominant and whose low cation exchange capacity has hydrogen as the dominant cation, possibly together with a certain amount of calcium. The generally flocculant character of these kaolin-type clay minerals will give rise to rather porous material whereas the great quantity of clay minerals will cause a strong development of planes of weakness. This causes a highly pedal fabric to develop which, due to intrapedal porosity, is reasonably friable. As was pointed out in Chapter 4, 1 : 1 clays are the most common product of the process of epimorphism; therefore, this type of porous highly pedal stable material formed from such 1 : 1 clays is also extremely common. Referring to Figure 4.4, materials necessary for this fabric would be produced in area B.

Fabric type (c) is developed from soil material in which 2 : 1 lattice clays are dominant and with a high cation exchange capacity, which is dominantly calcium. This is generally sufficient to maintain the clays in a reasonably flocculated state; because of this, a certain degree of porosity is given to the soil material. At the same time the high clay content ensures a strong pedal development. However, neither the porosity nor the stability of these peds is as great as that in type (b). Dark-coloured clay soils derived from basic rocks very commonly have this type of fabric (area C, Fig. 4.4).

Fabric type (d) is also developed from soil material in which 2 : 1 lattice clays are dominant but, as a result of suitable parent rock or site conditions, a considerable quantity of sodium and/or magnesium occurs as exchangeable cations. This is generally sufficient to cause easy deflocculation of the clay and leads to the development of dense non-porous material which, due to high clay content, still develops a considerable pedality. However, these peds are extremely hard when dry and sticky when wet – a great contrast with peds of types (b) and (c).

It must be emphasised that these four examples of soil fabric development are abstracted from a continuous spectrum resulting from a consideration of the condition of the matrix and the planar voids. Furthermore, by referring these fabrics to particular areas of Figure 4.4, there is no necessary implication of a time relationship between them, i.e. that with time, soil fabric will change from (d) to (c) to (b) to (a). The point that is being emphasised is that these fabrics are produced by a particular balance being achieved between all the factors involved in epimorphism. Fabric produced by the processes of lateral surface movement and by biospheric interactions will be discussed in subsequent chapters.

6

The process of lateral surface movement

Up to this point consideration has been given only to those processes concerned with the adjustment of lithospheric materials to the changed conditions of the surface. Another series of adjustments occurs at the surface of the lithosphere in response to a tendency to eliminate differences in gravitational potential consequent on topographic variation, namely the movement of lithospheric material from higher to lower topographic situations by interaction with elements of the atmosphere and hydrosphere. The overall process of movement consists of: (a) the erosion or detachment of particles from the surface; (b) the transport of such material; and (c) the deposition of the material. There are three agents involved in this threefold process: water, ice and wind. Each of these will be considered in turn.

The reactions of water with the surface of the lithosphere

There are two groups of reactions to consider: those in which water is the dominant factor and those in which it is only an accessory factor.

Reactions in which water is the dominant factor   This group of reactions includes those which are caused by rainfall, surface and subsurface flow. The physical effect of rainfall on the lithosphere surface is determined by its momentum which in turn is determined by raindrop size and velocity. Experimental results on single-grain materials of particular size have shown that rainfall as light as 1 mm/h contains drops which have sufficient momentum to disturb fine sand (Laws and Parsons 1943). Even though most materials at the Earth's surface are not single grain but are aggregated into more resistant entities, the fact that average raindrop size increases with increasing rainfall intensity means that particle detachment by rainfall on relatively unconsolidated materials is an important process.

On reaching the surface, rainfall is partitioned into surface and sub-surface flow. If the intensity of precipitation exceeds the rate of in-filtration, surface flow will occur; otherwise the entire rainfall will be absorbed and may become subsurface flow.

The nature of surface flow depends upon the effect of gravity, quantity of water and the character of the surface. The effect of gravity is a function of slope gradient; the quantity of water and roughness of the surface, together with gradient, largely determine flow velocity, depth and degree of turbulence (laminar surface flow is a rare occurrence under field conditions). It is to be expected that the erosive and detach-ing power of surface flow will depend upon its velocity which will gener-ally increase with distance from the crest of the slope, provided the slope is convex or straight. If it is concave, the decrease of gradient with distance may offset the effect of increasing quantity of water in a down-slope direction and erosive power could then decrease.

Moving downslope, surface flow becomes more concentrated into better defined and more permanent channels. Mechanical processes for channelled flow are essentially similar to those associated with more generalised surface flow, the chief point of contrast being in the spatial concentration of the channelled flow.

The degree to which moving water, as rain or surface flow, is ef-fective in detaching particles from a surface depends not only on what has been termed its **erosive power** but also upon the mechanical properties of the surface. Several properties are involved, the most important being cohesion between particles and particle weight. The cohesion between particles is largely determined by soil fabric and the factors involved in it. The inverse relationship of particle weight to ease of detachment is self evident and needs no elaboration except to note that the effect of weight will be modified by particle shape; generally, increasing sphericity means increasing ease of detachment.

Once a particle has been detached, it is likely to be transported downslope. Some of this movement is due directly to rainsplash, but the rates of such movement are not easily obtained (Carson and Kirkby 1972). In general it can be said that the distance over which a particle will be transported is a function of surface runoff velocity and of balance between the vertical component of turbulence and the settling velocity of particles, the last being influenced by the density and shape of the particle, as well as the roughness of the underlying surface.

As the flow of water becomes less turbulent, sedimentation will tend to involve those particles having a greater degree of sphericity and those which are nearer to the base of the flow. Since the processes of transport tend to sort particles according to their density, size and shape, deposits resulting from these processes will tend to contain particles of similar physical characteristics. In general, such sorting is greatest with channel flow, less marked with overland flow and least for rainsplash; also, the greater the distance of transport, the more precise the sorting.

The sorting resulting from lateral transport and deposition can be illustrated at various scales. Its importance with regard to materials found on lower hillslopes and neighbouring valley areas was emphasised by Milne (1947). Figure 6.1 shows the distribution of soils in the Unyanyembe catena in the Tabora region of Tanzania. The yellow–grey sand of the lower footslope and the sandy clay of the mbuga are explained as being due to the erosion of the skeletal grey loams of the hilltop and the red earths of the upper footslope. The quartz sand and the feldspar-derived clay which made up the red earth are sorted as they are eroded and transported downslope by running water. The quartz sand is redeposited on the lower footslope while the clay is either carried completely out of the system or deposited in the mbuga. A similar kind of redistribution of material according to the character of its ultimate particles is seen on a larger scale in the Kafue Basin of Zambia in Central Africa (Fig. 6.2). On the ridge area which fringes the basin in the east, epimorphic processes have operated continuously over a long period of time. Consequently, a bimodal particle-size distribution (clay–sand) is a common characteristic of the soil material. When such material is eroded and transported, no matter what the fabric, it will tend to be separated into its ultimate particles as it moves into the Kafue Basin. There is a relatively strong break of slope at the junction between the ridge and basin; at this point the quartz sand is deposited, forming an

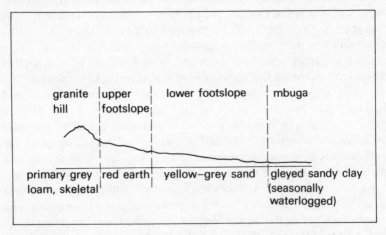

Figure 6.1 Soils of the Unyanyembe catena (after Milne 1947).

extensive and deep apron advancing from the margins towards the centre of the basin. The clay-sized particles are carried further downstream. The only exit for the water of the entire Kafue Basin is the very narrow gorge cut through the ridge. In times of flood, water banks up above the gorge and spreads over wide areas in the lower basin and the clays are then deposited over this flooded area, resulting in the formation of the heavy clay soils of the Kafue Flats immediately upstream from the gorge.

On a continental scale the same overall process can be recognised for it has led to the concentration of huge quartz sand deposits in the Congo and Kalahari Basins. While Africa provides probably the best examples of this process, it can also be recognised in Australia, India and South America in regions which have been land areas for an extended period of geological time.

The particles which form these sorted deposits give rise to fabrics that are very different from the four examples given in Chapter 5. At the depositional extreme the fabrics produced are the same as those considered by sedimentologists, but they have not adequately described

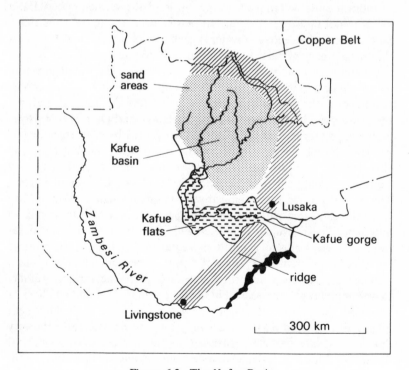

Figure 6.2 The Kafue Basin.

sand and clay fabrics developed under less than optimal fluviatile/ lacustrine conditions. In sands that result from transport, sorting and deposition the matrix fabric is non-uniform. It consists of areas, on a scale of a few millimetres, that are alternately tightly packed and highly porous; also the pores of such areas are highly irregular, both in size and form. Planar voids are absent or, at most, sporadically developed. The material as a whole is extremely fragile, easily breaking down to single-grain sands, for there is practically no finer-grained material to bind the quartz grains together.

The redeposition of clays permits closer packing between individual platelets or groups of platelets. This has a considerable effect on both aspects of fabric development. The matrix fabric becomes much more dense and, even though the planar void pattern remains quite well developed, its dominant mode is usually much coarser, while sub-dominant minor planar void development is much less than is the case with clays derived from bedrock by the operation of epimorphic processes.

We now turn to the question of subsurface flow. The rate and volume of infiltration is determined by the fabric of the surface material and by its initial moisture content. No matter what the initial conditions, however, subsurface flow of water is generally extremely slow. It follows that detachment of particles by subsurface flow is usually very limited and confined to the very finest of particle sizes. The initial moisture status could be important in certain situations for it has been shown that the drier the material, the finer the pores; and the faster the wetting, the greater will be the detachment (Panebokke and Quirk 1957). However, the greater part of the material transported by subsurface flow is probably by particles detached by rainfall impact. The fact that only clay-sized particles are readily mobile in such situations is illustrated by cases where only silt-sized particles are produced by rainfall impact. The silt is deposited almost immediately as a dense, hard-setting surface. Once the clay-sized particles are in suspension, they tend to remain so and deposition is usually associated with the loss of water by evaporation. This gives rise to depositional cutans of which a great number have been differentiated (Brewer 1964). The individual clay-sized particles are naturally deposited with their plate-like surfaces parallel to one another, giving rise to skin of clay orientated around any lines of weakness through which subsurface water can flow. Such movement affects all clay-sized particles, not only clay minerals but also the very fine ferric oxide particles generated in the process of new mineral formation. Thus, ferric oxide, despite being chemically immobile, is still capable of physical mobility under these conditions.

At this point it is worth noting that the mechanical and solutional modes of subsurface transport should be regarded as separate processes, even though they may sometimes operate simultaneously and in the same direction. Many writers have lumped the two processes under the single term 'leaching'. This is unfortunate. In this book, leaching describes subsurface solutional transport only (see Ch. 3) and another term to denote mechanical movement is required. **Pervection,** from the Latin *pervehere* meaning 'convey through', is proposed and will be used from this point onwards when referring specifically to the mechanical movement of solid particles by subsurface flow.

The way in which water reacts with soil material is greatly affected by the amount of dissolved ions which the water contains. These ions come from the oceans and the land areas of the world, mainly via the atmosphere, and are returned to the Earth's surface by rainfall. Chloride, sulphate, sodium and magnesium from the oceans, with calcium and potassium from the land, are dominant (Hutton 1958; Erikson 1959 and 1960; Gorham 1961). The ionic content of water at the lithosphere surface varies continuously, depending on the supply of ions from the processes of weathering and leaching. The presence of these ions should be taken into account for any study of flocculation and deflocculation of clays and, consequently, their effect on soil fabric as well as detatchment, transport, deposition and pervection, particularly in coastal zones and saline basins of inland drainage. This point has been emphasised by Mattson *et al.* (1944), Gibbs (1949) and Carroll (1962), even though the relationships remain highly speculative.

Reactions in which water generally has an accessory role In all the processes considered so far water has been the dominant element, the solid material accompanying the water in a rather inert manner. On the other hand, there is another group of processes in which the solid material reacts to gravity with water playing no more than an accessory role. Movements dominated by gravity range from imperceptibly slow to very rapid.

Creep is the term generally applied to the slow process of movement extending over long time periods (Young 1972). This movement is regarded as being quasi-continuous, but the actual mechanics of creep are still obscure. The more rapid movements of surface material include a great range of processes from landslides, where the movement of material occurs en bloc after failure along one or more discrete surfaces, to mudflows where the moving mass has a higher water content and often follows former stream courses (Sharpe 1938).

In terms of surface movement as it affects the formation of soil

material, the chief characteristics of landslides and mudflows are:
(a) they are intermittent and individually localised; (b) their deposits
are initially unconsolidated and unsorted, although such deposits may
thereafter be rapidly sorted by water or wind; (c) previously sorted
material becomes less well sorted and loses, either in part or as a whole,
its former stratification and fabric; and (d) fresh material is exposed at
the site of movement so that rejuvenation of weathering, leaching and
new mineral formation occurs. Our knowledge of soil-forming processes
and materials on surfaces affected by rapid mass movement is very
limited; however, the importance and widespread occurrence of such
movements, so far greatly underestimated, need to be stressed for
any realistic study of hillslope soils. In areas where mass movements are,
or have been, numerous, notably those of high relief but also any
locality where slopes are steep enough to be potentially unstable, the
distribution of superficial materials is likely to be complex; they will
vary in condition, fabric and degree of epimorphism. Needless to say,
this has significant implications for soil mapping, typology and classification.

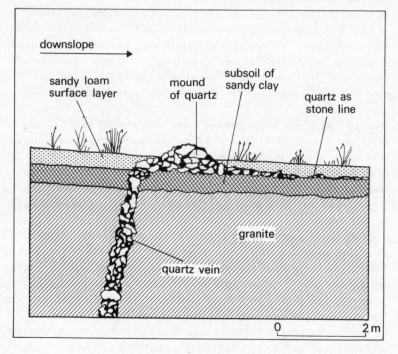

Figure 6.3 Quartz vein on hillslope.

The nature of soil material on hillslopes: four case studies

In order to treat the formation of soil material on hillslopes in a more integrated manner, a stratigraphic approach appears to be essential. To illustrate this point, four case studies have been chosen from fieldwork recently undertaken in eastern Australia. Figure 6.3 shows the soil material that occurs on a gentle hillslope in the Brisbane Valley of southern Queensland, where the bedrock is granite. A sandy loam surface overlies a sandy clay subsoil which grades into the underlying granite. The relationship of these two layers is made clear by the way in which the quartz vein behaves as it approaches the surface. It passes through the sandy clay of the subsoil without disturbance, but when it reaches the sandy loam surface it forms a mound just downslope from the outcrop. Downslope from the mound the quartz fragments occur as a distinct stone line at the junction between the surface layer and the clay subsoil. No stone line exists between the two layers upslope from the point of outcrop. This can be interpreted as showing that the sandy loam topsoil is moving downslope across the sandy clay subsoil, which is being formed by *in situ* alteration of the granite. The lateral continuity of the surface layer over considerable segments of the landscape suggests that movement has been slow, relatively continuous and possibly contemporary.

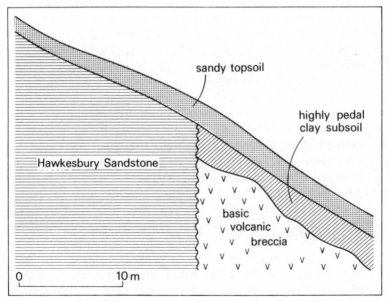

Figure 6.4 Hillslope across boundary of diatreme.

Even stronger evidence for the downhill movement of a lighter-textured surface layer is to be found in the Sydney Basin where Hawkesbury Sandstone has been penetrated by diatremes filled with brecciated basic igneous rocks. Because they can be eroded more easily than sandstone, the breccias occur in depressed sites surrounded by steep slopes at the sandstone boundary. Figure 6.4 is typical of the types of soil material developed across such a boundary in the Hornsby area of Sydney. On the upper slope there is a uniform sand directly overlying the sandstone but, immediately the boundary with the breccia is crossed, a highly pedal clay develops under the surface layer of sand, giving rise to a typical texture-contrast soil. In this case, downslope movement of the surface sand can be established by an examination of the nature of the quartz sand grains. Quartz grains with well-terminated pyramidal faces are a common feature of the Hawkesbury Sandstone and can be clearly recognised within the surface sandy layer, no matter whether it overlies the sandstone directly or the clay subsoil. However, there is no sign of this type of quartz grain within the highly pedal clay material overlying the diatreme breccia. Similar evidence is provided by differences in heavy mineral separates. The sandy surface contains ilmenite, rutile and zircon, both where it directly overlies the sandstone and where it occurs on the clay subsoil; these minerals are typical of the Hawkesbury Sandstone but they are completely absent from the clay subsoil which contains strongly altered olivine derived from the breccia. It can be concluded that the clay subsoil has been formed by the *in situ* alteration of the breccia and that the sandy topsoil is derived from the Hawkesbury Sandstone. The surface layer of sand has moved and is still moving downslope across the top of the breccia-derived clay.

These two examples are taken from sites where there is a strong texture contrast between the surface and the subsoil, but on the Darling Downs in southern Queensland there is similar evidence of downslope movement in soil materials which have a clay texture throughout. In general, the landscape consists of flat-topped residuals surrounded by long gentle slopes. The basement rocks are a variety of basaltic materials, both lavas and ashes, which are altered to varying degrees. At the upper end of the long, gentle hillslopes a shallow, gravelly, dark clay occurs (Fig. 6.5). The dark clay appears to pass over the varied basement without any alteration in its character. However, examination of the gravels shows that they are derived from each of these different basaltic materials; each gravel can be traced for a short distance downslope from the point of outcrop of its particular basaltic source. This is particularly well seen in the case of the red-bole layer. The fact that the gravels can be followed only in a downslope direction is strong evidence

University of Chester, Seaborne Library

Title: Geoforensics / Alastair Ruffell and
Jennifer McKinley.
ID: 36064373
Due: 06-05-10

Title: Forensic entomology : an introduction /
Dorothy E. Gennard.
ID: 36078667
Due: 06-05-10

Title: Geological and soil evidence : forensic
applications / Kenneth Pye.
ID: 36055392
Due: 06-05-10

Title: Soils / William Dubbin.
ID: 36084378
Due: 06-05-10

Title: The formation of soil material
ID: 78166001
Due: 06-05-10

Total items: 5
15/04/2010 15:31

Thank you for using Self Check

3

that downslope movement of the surface dark clay is taking place.
Furthermore, it can be inferred that movement is taking place at a slow
rate for, although it takes some time for the basaltic gravels to lose their
distinctive character, they do so within a few feet of their point of origin.

These three examples show that slow continuous downhill movement
of material is an important contemporary process. There is, however,
no evidence to indicate whether movement is faster at the surface than
at the base of the mobile layer or is uniform throughout. Nor is it
possible to decide whether this movement is achieved by creep or wash,
or some combination of the two. However, a detailed investigation of
soil fabric may well lead to a solution of this problem. The stratigraphic
approach also reveals the extent to which the formation of soil material
is affected by sequences of erosion and deposition that may not be very
evident from a superficial examination of hillslope profiles.

In the Darling Downs example (Fig. 6.5) only the very top of the
hillslope was considered, where the dark clay directly overlies the
bedrock. Moving further down the slope (Fig. 6.6), a yellowish–brown
clay with carbonate nodules appears beneath the surface layer, at first
as isolated patches in bedrock hollows and then further downslope as a

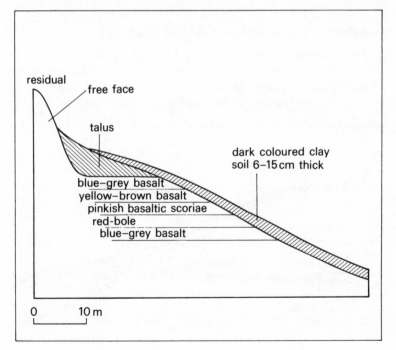

Figure 6.5 Hillslope on varied basaltic materials.

continuous layer. This clay is in turn underlain, slightly further down-slope, by a reddish–brown clay with abundant soft carbonate. Such an arrangement of materials means that periods of erosion, transport and deposition affected this slope before the continuous downslope movement of the present surface layer was initiated. Another example is provided by the long, gentle hillslope forming the eastern side of Tenthill Creek Valley forty miles west of Brisbane in southern Queensland (Fig. 6.7). The basement rock is a Mesozoic sandstone which is shown approaching the surface near the top of the hillslope. Between the sandstone and the continuous surface layer (1) there are three distinct bodies of sandy clay material which are numbered (2), (3) and (4) from youngest to oldest. The cut-and-fill relationship which obtains between these materials indicates that erosion and deposition have been discontinuous both in space and time and have played a major role in determining the character of soil-forming material on the slope. The sporadic episodes of cut-and-fill of the two examples may be related to periods of landscape instability, while the contemporary continuous movement of the surface layer is associated with a period of landscape stability.

The action of ice at the surface of the lithosphere

Ice action is most conveniently considered in the order: erosion or detachment, and then transportation and deposition. The detachment of particles can be caused by two distinct processes: the expansion of ice when formed from water, referred to as **frost shattering,** and the

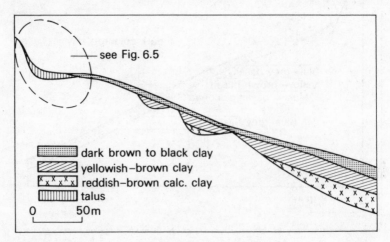

Figure 6.6   Extended section of hillslope on varied basaltic materials.

movement of ice across the lithosphere surface. The process of frost shattering can occur only with the right temperature variations and this generally means high altitudes or high latitudes, or both, although sometimes the right conditions occur nearer the equator under desert climates. To be effective as a disruptive agent the expansion associated with ice formation must occur within material that is comparatively incompressible. Thus, in general, soil material acts as a blanket against the disruptive action of ice formation. It is only in certain heavy clay soils where water is confined to very fine planes of weakness that frost shattering does occur and it operates in this case due to Man's intervention by ploughing only (Russell 1961). Apart from such exceptions, frost shattering is confined to relatively fresh bedrock, particularly bedrock with a well-developed system of joints. The effectiveness of this process in supplying coarse fragments is probably of considerable significance for the actual expansive pressures involved are very marked. This is in contrast to the so-called 'insolation weathering', which suggests that rocks and minerals are ruptured as a result of large diurnal and/or seasonal temperature changes, where the actual forces involved are minute. It would now seem that for 'insolation weathering' to occur it must be preceded by a certain amount of weathering, leaching and new mineral formation (Birot 1968). However, this would not seem to be the case in frost shattering.

Ice moving across bedrock has a grinding action largely due to the rock material transported at the base of the ice. The process of grinding

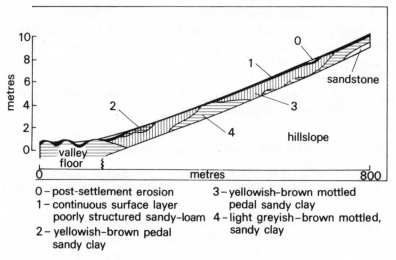

Figure 6.7 Hillslope on varied sedimentary rocks.

is responsible for producing a whole range of particle sizes, from sand to fine clay. This material differs profoundly from that resulting from the action of weathering, leaching and new mineral formation, in that there is no bimodal distribution of particle size, and silt-size particles are of common occurrence. Furthermore, the particles resulting from the action of weathering, leaching and new mineral formation tend to be rather inert, for the quartze sand grains are residual and the newly formed clay-size particles are in balance with their environment. The particles resulting from ice grinding, however, consist of shattered silicate structures very much out of balance with their environment and hence extremely reactive. These two unique characters of material produced by ice grinding have profound implications in terms of the subsequent action of sorting processes and the soil material produced.

The transportation and deposition of material by ice can be dealt with relatively briefly. No matter how it is transported – whether on, within or at the base of the ice – there is little differential movement of the material while it is associated with the ice and hence deposits are a mass of unsorted and unconsolidated material. In fact, the overall effect of ice action is very similar to that of landslides and mudflows, for in the process of detachment, particularly frost shattering, much more material is exposed to the action of surface alteration and the unconsolidated, unsorted deposits are available for the accelerated action of water and wind. This similarity is even more marked in the case of the process of **solifluction,** which occurs within regions of continuous permafrost where summer thawing is limited to a shallow depth. Thaw water cannot penetrate into the permafrost beneath and so it remains in the surface layer, saturating it and permitting rapid flow of that layer downslope, on slopes as gentle as 2°. As a result, slopes can quickly become mantled with this unsorted material within a relatively short time period (Flint 1971). However, owing to the unique character of the material involved, the sorting processes of wind and water produce some equally unique deposits.

The action of wind at the surface of the lithosphere

The action of wind in detaching particles from rock is dependent upon its own content of solids in much the same way as the erosive action of ice is dependent upon the rock fragments it contains. However, its ability to erode consolidated materials is very limited and wind erosion is only significant for unconsolidated, sparsely vegetated and relatively fine-grained materials. It is very rare for any particles bigger than sand size (2 mm) to be moved by wind across a rough surface. Particles between 2 mm and 0·05 mm generally move by saltation, while particles less than 0·05 mm travel in suspension for they can be moved by eddies and, if very fine, diffused to great heights in the atmosphere. This means that finer particles are carried away more quickly and are clearly separated from coarser particles. However, the entrainment of finer material is dependent upon the presence of particles of varied size in the source material. In a well-sorted, fine-grained deposit particles are not easily entrained because surface roughness is slight, there is significant inter-particle cohesion and, most importantly, in the absence of coarser particles, there is no saltation to initiate particle movement by bombardment.

In general terms, the products of wind action depend upon a combination of three factors: the duration of epimorphism, the previous detachment and transport of particles by water or ice, and an arid environment. In hot deserts where, typically, prolonged epimorphism has generated (as explained in Ch. 4) a bimodal distribution of particle sizes, both sand and clay-sized particles will be available for removal, chiefly as distinct bodies of water-sorted sands and clays. Deflocculated clay particles will be removed in suspension, carried high into the atmosphere and transported considerable distances before being redeposited. This is readily apparent in the dust storms which are such a common phenomenon in areas surrounding deserts. Such dust may be of considerable pedological importance for, although the amount deposited from individual storms is very small in terms of depth per unit area, the accumulation over a long period of time may be far from negligible. However, it is difficult to determine the amount of material involved in this process, for not only is the deposition difficult to measure, but the particles are usually incorporated into underlying soil material where they are likely to be redistributed by pervection. At present their provenance is a matter for inference rather than direct identification, although development of techniques of determining oxygen isotope ratios (Taylor and Epstein 1962), which for any particular mineral such as quartz are characteristic of the conditions under which the mineral formed, offers the possibility in the near future of determining even a relatively small wind-added component in a particular soil material. Discrete bodies of airborne clay particles only form under exceptional circumstances; e.g. if the clay is flocculated into sand-sized aggregates, clay dunes may form in the same way as sand dunes, as exemplified by the clay 'lunettes' downwind of desiccated clay pans in Australia (Bowler 1973).

Materials in the range 2–0·05 mm move by saltation, leaving behind any coarser particles, and tend to assume various dune forms. The presence of modern active dunes is easily recognised but there is much evidence for the widespread occurrence of ancient aeolian sand dunes, now stabilised by vegetation, in semi-arid and coastal zones as well as, more locally, in humid environments affected by aridity from time to time during the Quaternary period.

In the cold, arid environments marginal to the Pleistocene ice sheets, initially unsorted morainic material contained a significant proportion of silt-sized rock debris. Although the precise mode of silt movement in air has received little attention, its grain size suggests movement in suspension but, since its settling velocity is relatively great, silt is dispersed less widely than deflocculated clay. This, coupled with the fact that large quantities of silt became available for aeolian removal in a comparatively brief period during the Quaternary deglaciations, led to deep deposits of material (loess) in many periglacial areas. In such blanket deposits the silt particles are associated with each other in a highly porous 'loess fabric' very similar in character to the highly porous earthy fabric described in Chapter 5 (Fig. 5.5). It is likely that silt-sized particles were also deposited beyond these distinctive loess sediments as thin veneers to act as a modifying influence on pre-existing soil materials; this possibility requires further investigation.

Conclusion

In considering the processes of lateral surface movement a very broad view has necessarily been taken. The importance of these processes for the formation of soil material has been illustrated and will receive further emphasis when discussing the concept of pedological provinces in Chapter 10. To conclude this chapter it is worth remarking that the significance of lateral surface movement has hitherto been under-estimated. This partly, perhaps primarily, stems from the lack of an integrated approach to the subject consequent upon the linkage between the separate investigations undertaken within the fields of pedology and geomorphology. This is particularly evident with regard to the study of processes and materials occurring on hillslopes. Geomorphologists have concentrated on slope form and the downhill movement of material, paying only passing attention to the nature of the soil material as such. On the other hand, pedologists largely confine themselves to profiles and continue to regard lateral processes of surface movement as non-pedological. The four case studies considered earlier in this chapter (p. 61) demonstrate that any view of soil formation which excludes or ignores what may be termed **geomorphic processes,** on the grounds that they are not pedological, is too narrow an approach to soil formation and is bound to lead to erroneous conclusions.

7

The effect of the biosphere on the processes of epimorphism

Up to this point the model of soil formation has been developed without considering the way in which the biosphere reacts with soil material. As the biosphere is concentrated to a great extent along the lithosphere–atmosphere interface it follows that the number of reactions between it and soil material are numerous and extremely complex. Taking an overall view, however, it is possible to rationalise this complex in terms of the model of soil formation by considering biospheric reactions under two main headings: reactions with epimorphic processes, i.e. weathering, leaching and new mineral formation, which will be considered in this chapter; and reactions with the processes involved in the lateral movement of material across the surface of the lithosphere, which will be dealt with in the next chapter.

Weathering as affected by the biosphere

As stated previously, weathering is the breakdown of complex silicates into simpler entities. The biosphere is involved in this process both directly and indirectly. The main process involved in weathering is hydrolysis, the speed of which is dependent upon the hydrogen ion concentration. Therefore, biospheric processes which increase the hydrogen ion concentration will indirectly increase the speed of weathering. Living roots are of particular importance for, by reacting with their environment to obtain the nutrient elements essential for their existence, they increase the hydrogen ion concentration of their surroundings. This is mainly by excreting carbon dioxide which forms a solution of carbonic acid on the surface of minerals. Hydrogen ions are then formed by ionisation. Secondly, there is a process of direct exchange of hydrogen ions excreted by the roots with cations held by the mineral particles. Roots may also be capable of excreting organic acids (Russell 1961). The breakdown of organic matter also produces organic acids at intermediate stages, while the final breakdown product is carbon dioxide. Thus, both living and dead organic matter is capable of increasing the speed of weathering indirectly.

It is also possible for organic matter to participate directly in the weathering process by forming complexes with metals in various ways, including chelation. Chelation refers to a reaction between a metal ion and a complexing agent and is characterised by the formation of a ring structure. Insofar as the formation of these complexes requires the removal of metals from the silicate structures in which they originally occurred, the organic matter involved can be regarded as a weathering agent. Such a process can be regarded as a unique biospheric contribution to the process of weathering.

Organic molecules capable of forming complexes occur in both living and dead materials. It seems impossible to explain how plants obtain the iron necessary for their survival other than by the excretion of complexing agents by plant roots as, in normally aerated soils, ferric iron has such a low solubility and ferrous iron cannot be maintained at a sufficient concentration (Wallace 1963). There is also some evidence to suggest that phosphate supply to certain plants is secured in a similar manner (Russell 1961). In the case of soil organic matter, Beckwith (1955) found evidence that it formed complexes with divalent metal cations, such as copper and manganese, and suggested that in the case of trivalent ions, such as ferric iron and aluminium, much stronger complexes could possibly be formed. Increase in hydrogen ion concentration and the direct complexing action of organic matter are both characteristic of all materials making up the biosphere. However, it seems that certain groups of organisms have a greater potential than others for utilising these two mechanisms. Thus Jacks (1953), quoting Russian work on the initial colonisation of bare rock surfaces by lichens, showed that they had a much greater ability than more highly evolved plants to extract elements such as potassium and phosphorus essential to their continued existence. Again, diatoms require silica for their skeleton; to acquire this they must be capable of breaking down the framework of silicate minerals. Consequently, such organisms must be very effective weathering agents.

Leaching as affected by the biosphere

Leaching is the movement of simpler entities produced by weathering.

Insofar as such material is removed by the growth of organisms, it is retained within the biosphere and leaching is retarded. On the other hand, organic matter moves into the surface of the lithosphere and reacts with it to form organo-metal complexes. Some of these are highly insoluble and, again, leaching is retarded. In other cases soluble complexes are formed, in which case the process of leaching is accelerated.

Retardation Plants generally absorb more of certain elements than of others. Thus, potassium appears to move more quickly from the soil into the plant than do other common cations and nitrate more than other anions, so that these two ions appear to be taken up in appreciably higher proportion than their relative abundance in the soil would suggest (Russell 1961). Referring back to Figure 3.1 (the ionic potential diagram), it can be seen that potassium belongs to the simple cations of group I, while nitrate belongs to the complex ions of group III. Hutchinson (1943) plotted the concentration of elements in terrestrial plants against their concentration in the lithosphere on the same ionic diagram; the result is shown in Figure 7.1. From this it is

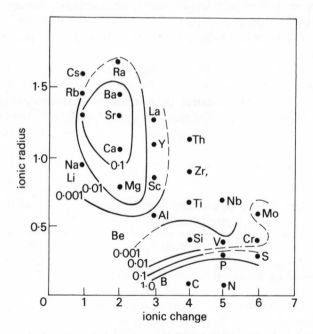

Figure 7.1 Concentration ratio of elements in terrestrial plants (after Hutchinson 1943, by permission of Williams and Wilkins).

apparent that, in comparison with the elements of group II, the more mobile elements of groups I and III are relatively concentrated in terrestrial plants. These are, of course, exactly the elements which are most subject to leaching and so the retarding effect of the biosphere on their removal is all the more marked. However, this does not imply that the more immobile but common elements of group II are not concentrated by particular plant species. Lovering (1959) quoted examples of grasses accumulating up to 1 % of their dry weight as silica, while sugar cane was shown to be capable of removing 1900 kg of silica per hectare in a two-year growth period. Many tropical trees and some temperate zone hardwoods were also shown to be silica accumulators.

Despite the generally low content of aluminium in terrestrial vegetation (Fig. 7.1) it has been shown by Hutchinson (1943) and Chenery (1948) that particular plants are capable of accumulating aluminium to a very high level. The most remarkable example is that of the Australian silky oak (*Orites excelsa*) which actually secretes aluminium succinate in cavities in the wood; its ash is 75 % Al_2O_3. Of more than 4000 plant species investigated by Chenery (1948), almost half were found to accumulate aluminium to a certain extent, while only one species was found to be an iron accumulator. This seems to confirm the inert character of ferric iron which is a marked feature of other phases of epimorphism.

However, with respect to a great range of elements present in small amounts and in a highly dispersed state in the lithosphere, plants appear to act as general concentrators, no matter to what ionic potential group the elements belong. Goldschmidt (1937), working on material from old beech and oak forests growing on sandy soils in Germany, showed that fresh oak leaves concentrated boron over 1500 times, manganese fifty times and nickel twice relative to the concentration in

Table 7.1 Enrichment of elements during decay of oak and beech humus (in ppm) (after Goldschmidt 1937).

	B_2O_3	MnO	NiO	GeO_2
Mineral soil (sand)	7	400	20	5
Ash from fresh oak leaves	5–10 000	20 000	50	5
Ash from oak humus	200	2400	100	70
Ash from beech humus	30	1400	100	70

Other elements concentrated in the humus – As, Ag, Au, Be, Co, Zn, Cd, Sn, Pd, Ti the soil (Table 7.1). The degree to which these materials are retained in the organic-rich surface horizon of the soil varies. The relative concentration of both boron and manganese drops considerably, while that of nickel doubles and that of germanium increases fourteen times. Other elements concentrated in the organic-rich layers of the soil are As, Ag, Au, Be, Co, Zn, Cd, Sn, Pd and Ti. It could be that a certain amount of this concentration is due to the formation of insoluble organo-metal complexes, as suggested by Beckwith (1955).

Both specific and general element concentration by terrestrial vegetation has been demonstrated. Naturally, this is of fundamental importance to the biosphere and biospheric development, but what of its importance in the model of soil development now being considered? Is it possible to determine the amount retained by the biosphere compared with the total quantity of material involved in the process of epimorphism?

It has already been stated that the initial colonisers of bare rock surfaces, such as lichens, are responsible for an intensification of weathering. At the same time they are responsible for a relative concentration of calcium, magnesium, potassium, iron, phosphorus and sulphur within themselves, compared with that in the underlying rock. Once such a concentration has been achieved it tends to stay in the biosphere, for these elements are cycled from one generation of plants to the next by way of litter fall. Also, more of these elements continue to be concentrated within the biosphere as long as the feeding roots of the vegetation penetrate to the weathering zone. When this no longer occurs, the only source of nutrients for continued plant growth is the litter fall of the previous generation of plants. Given that the thickness of the weathering mantle will increase with time, there is a tendency for the quantity of material of lithospheric origin in the biosphere to approach a limiting value. If such a limiting situation containing the maximum biomass to be found on the surface of the earth could be located, it would then be possible to determine the maximum amount of lithospheric elements retained by terrestrial vegetation. However, there is a great lack of basic data in this area, most investigations dealing only with a limited number of elements. Nye and Greenland (1960) determined the cycling of the major nutrient elements (calcium, magnesium, potassium, phosphorus and nitrogen) between various secondary vegetational types and the soil, but did not take account of elements such as silicon, aluminium and iron. Even the comprehensive account of mineral cycling in the biosphere by Rodin and Bazilevich (1965) serves to emphasise the paucity of basic facts. However, taking the mineral content of the tropical forest (the maximum biomass) that these writers give and comparing this content with the amount of elements contained in a 10 cm thickness of average granite, it can be shown that, even in the case of potassium and phosphorus, representative of the mobile elements of groups I and III, total retention in the biomass would not exceed 2 to 3%, while in the case of the more immobile elements of group II, such as silicon and aluminium, the total amount retained is very much less. Thus, in the overall scheme of epimorphism this aspect of the part played by terrestrial vegetation should not be overemphasised.

Acceleration   In dealing with the question of the acceleration of the process of leaching, a much more restricted view is taken than in the previous section. In this case consideration is being given only to the very special circumstances in which iron and to a lesser extent aluminium become very much more mobile because of reactions with organic molecules (Crompton 1962). In normal soil materials the highly active organic molecules produced by the breakdown of materials from plants and animals are inactivated by reaction with clay mineral particles. However, as a result of the processes of lateral movement, there are a considerable number of depositional sites at which quartz sands are segregated and accumulate. In such materials there are no clay minerals to inactivate the organic molecules and the only materials with which it is possible for the organic molecules to react are the fine-grained iron oxides which very commonly form a skin around the quartz grains. The organic molecules form a chelation compound with the iron which, as a result, is removed from the quartz grains, giving them a bleached appearance. The iron is moved in the chelate form deeper into the quartz sand where it is finally redeposited as a pan, i.e. a podzol is formed.

Bloomfield (1954) indicated that the active organic molecules were simple polyphenols produced by a whole range of plant materials, while Handley (1954) showed that the simple polyphenol content in plants is related to the nutrient status of the soil material. The poorer and more acidic the material (i.e. the more quartz-rich), the higher the content of simple polyphenols. Coulson *et al.* (1960) and Davies *et al.* (1964) showed that the maximum quantity of simple polyphenols occurred in freshly growing green leaves, with much less in senescent or dead leaves and minimum amounts in the humus layer on the soil surface.

It should be noted particularly that the production of podzols is dependent upon some very special circumstances. Therefore, the usual acceptance of podzolisation as one of the general pedological processes is not justified and if so accepted is likely to give rise to some very erroneous conclusions.

New mineral formation as affected by the biosphere

In general, the silicate framework fragments, which go to form the new minerals, are highly reactive; the same can be said of the organic breakdown products. The way in which these two kinds of entities interact to form the organo-mineral complex of topsoils is still to a large extent unknown because of the great difficulties inherent in its investigation. As far as the organic molecules are concerned, there is still considerable uncertainty as to the actual entities involved, and it is only since more sophisticated forms of analysis became available that any real progress has been made (Greenland 1965a and b). The inorganic fraction presents just as many unknowns for, while types of crystalline and amorphous material have been recognised in soil, little is known about how these materials are actually arranged in particular soils. Little is therefore known of the type of surfaces with which the organic molecules come into contact, but they could be composed of silicates, oxides and hydroxides with varying degrees of internal order and perhaps coated to a greater or lesser degree with amorphous materials.

The special circumstances of the North Island of New Zealand provide an example where these reactions have been evaluated. In the chapter on new mineral formation (Ch. 4), it was shown that in the sequence of soils derived from volcanic ash there is a very rapid sequence of weathering, leaching and new mineral formation, giving rise first to allophane and then to metahalloysite as the dominant clay minerals. This rapidity of development permits evaluation of organic–mineral interactions in the surface horizons of such soils. It was found (Fieldes 1955) that, in the allophane of these surface horizons, the aluminium and silicon occur in discrete lattices whereas, away from the surface, both silicon and aluminium are normally in the same lattice. The abnormal type is termed **allophane B**, the more normal type **allophane A.** In allophane B, organic matter forms a complex with aluminium which prevents its mutual precipitation with silicon to form the more normal allophane A. This inhibits the further progress towards metahalloysite which would occur in subsoils away from the blocking influence of organic matter.

In many cases, however, the reactions are deduced from general chemical effects. Thus the removal of interlayer potassium is the essential process in the transformation of primary micas to illites and clay vermiculites. Since, as a result of biospheric activity, there is a considerable concentration of potassium in the surface zone, it may become great enough to offset the general trend towards potassium removal causing illite to develop in the surface zone from clay-vermiculite by the accommodation of potassium in interlayer 12-co-ordination (Fieldes and Swindale 1954). The ammonium ion is also produced in this surface zone and is sufficiently near the size of potassium to react in the same way, as has been shown experimentally (Russell 1961). In other cases these reactions are postulated on very generalised correlations, as was done by Jacks (1953) reporting Russian work on the formation of montmorillonite as a result of the colonisation of bare rock surfaces by lichens. This was ascribed to the general environment being rich in metal cations because of biospheric activity.

Fabric as affected by the biosphere

Despite the lack of knowledge regarding the way in which organic and inorganic entities react, there is no disputing the fact that they do and that, in the process, they mutually inactivate each other. This results in the organo-mineral complex in which the normal bonds within the soil material are strengthened. Emerson (1959) showed that some types of organic matter strengthen the normal bonds between clay minerals and quartz and so would stabilise any porosity already present in the soil material. However, the evidence is presented in terms not of porosity, but of pedality; in these terms Emerson showed that organic matter is able to increase crumb stability. Field observations confirm these experimental results but it is difficult to say whether the structure is due to the organic molecules as such, to the living roots of the plants, or to the soil fauna; it is probably a combination of all three and most interactions involving the biosphere must be viewed in this way. The effects of termites, ants and worms will be discussed in the next chapter.

In contrast, reactions associated with the simple polyphenols produced under the specialised conditions necessary for podzol formation are of a very different nature. Initially, the sand body in which the podzol develops would have the mixed, closely packed and highly porous fabric with few planar voids described in Chapter 6. The stripping of the ferric oxides from the quartz grains and the removal of any finer grained material result in a single grain sand, while the deposition of iron and organic matter at greater depth produces the massive, dense, indurated and brittle pan fabric.

8

The effect of the biosphere on the processes of lateral movement

Interactions between plants and processes of lateral movement

In general, vegetation slows down the processes of detachment, transportation and deposition insofar as they are controlled by water and wind, but is without effect in the case of ice action and the larger landslides and mudflows.

Plant cover invariably reduces the erosive potential of raindrops by interception; in the case of a continuous canopy relatively near the ground, rainsplash erosion may be reduced to negligible proportions. Similarly, vegetation tends to reduce sheet erosion, particularly where it consists of a mass of fine roots and stems. Not only is the surface protected from scour but plant roots increase the mechanical stability of the soil and reduce the extremes of wetting and drying and also the speed of wetting. Plant roots are also undoubtedly responsible for slowing down any downhill movement that can be ascribed to creep. Vegetation virtually prevents the wind having any effect on the lithosphere surface in terms of detachment, entrainment and transportation of material. In terms of deposition, vegetation generally provides a more stable site around which wind-transported material will accumulate.

It is of interest in the perspective of geological time that this control of surface processes by vegetation started only in the later Silurian with the initial development of terrestrial vegetation (Berkner and Marshall 1965) and only with the development of grasses in the Cretaceous has the present vegetational control been in operation. Thus, the generally conservative action of vegetation has been operative for only 100 million years and as far back as 400 million years to a lesser extent.

Prior to that, for a period of at least 2600 million years, the processes of lateral surface movement must have operated at much greater speeds and much more completely in terms of end products, unhindered by any vegetation.

Despite the fact that most plant–lithosphere interactions slow down the processes of lateral movement, there are a few which accelerate the rate at which these processes operate. The most commonly quoted instance is the detachment of bedrock fragments by the expansive power of tree roots in the course of their growth. The reality of such pressures and the speed with which they operate is to be seen in the common occurrence of cracked roadways and footpaths adjacent to tree roots. Given such a penetration of roots through the soil and into the bedrock it follows that, if tree-fall occurs, there must be a drastic disturbance of both soil material and bedrock. Such disturbance was recognised from an early date (Shaler 1890–91, Van Hise 1904), but quantitative data on this topic remain sparse. Lutz and Griswold (1939) investigated the detailed morphology of soils within the Yale Research Forest in New Hampshire. In a series of sections across wind-throw features, they showed that the soil profile had been so profoundly disturbed to a depth of about 1 m that the relationship between soil materials within the forest could not be understood without taking this process into account.

Stephens (1956), working in the Harvard Forest, was able to show that in an area of one acre almost 14 per cent of the surface was composed of moulds and pits characteristic of tree-fall. Cross-sections of these features confirmed the extreme disturbance of the soil material. Stephens further suggested that this amount of disturbance had taken place in about 500 years. With this degree of perspective, tree-fall assumes the proportions of an important pedological process, particularly as these detailed observations appear to be applicable to the whole of the east coast of North America.

The author's personal observation suggests that tree-fall as a pedological process is also important in equatorial lowland dipterocarp forest. On the east coast of Sabah, Malaysia, a 10 m high coastal platform has two contrasted soils on its surface: a shallow one up to 1 m deep overlying steeply dipping Miocene sandstones across a sharp boundary and a much deeper soil on unconsolidated Quaternary clays and sands. Tree-fall is much more common on the shallow soil than on the deeper one, this being a reflection of the depth of rooting. This difference can be recognised on aerial photographs in terms of continuity of canopy or the lack of it, which must mean that the rate at which tree-fall occurs is considerable. When it is realised that each tree-fall entangles within its root mat a mass of material 3–4 m in diameter and 20–40 cm thick, its pedological significance is readily apparent.

Another positive effect of vegetation on the processes of lateral movement is to be seen when plants and their decay products are involved in bog-flows or bog-bursts, which involve the downhill movement of masses of water-saturated peat, common in the mountainous areas of northwestern Europe, particularly Ireland. The largest of these flows, involving up to 5 000 000 m^3 of peat, is associated with raised bogs, in which a domed mass of soft peat is enclosed within a perimeter of firmer material. On receiving a sudden addition of water, swelling occurs until failure at some point on the perimeter releases the semi-fluid peat. More limited peat-flows are associated with blanket bogs developed on hillslopes up to a limit of 15°. As the slope increases, the volume of material involved decreases and at the same time the nature of the movement changes from a flow into something which approximates a slide. As in the case of other flows and slides, such movement of organic material accelerates the processes of epimorphism by the removal of the protective mantle.

Interactions between animals and processes of lateral movement

Animal control of vegetational cover   Insofar as animals reduce the vegetational cover, they reduce its overall conservative action on the processes of lateral movement. This is reflected in the numerous reports on the effects of heavy grazing. An example is provided by the subsurface mudflows, reported from southeastern Australia, where the trigger mechanism for the whole process is overgrazing by stock (Downes 1946).

Giardino (1974) observed a great increase in the amount of surface runoff and gullying within the Luangwe National Park, Zambia, paralleling an increase in the elephant population caused by protection and a decrease in the amount of vegetation, for each elephant consumes about 200 kg of vegetation daily and destroys even more in the process of obtaining the food.

The activities of harvester ants and termites are responsible for the removal of considerable quantities of grass cover which, in certain semi-arid areas where the initial cover is not too abundant, increases the speed at which the processes of detachment, transportation and deposition occur. Thorp (1949) described such an effect from the Great Plains of North America, where between ten and twenty ant-hills per acre occur and where each one is surrounded by a circle of bare soil

2–10 m in diameter. In addition, certain of these harvester ants construct paths, 10–20 cm wide and 20–30 m long, that radiate out into the surrounding vegetation (Wheeler 1910).

Termites are equally effective; Watson and Gay (1970), working in southwestern Queensland, found that a succession of favourable seasons had promoted a prolific growth of native grasses which, in turn, had caused the termite population to increase. In ensuing seasons of poor grass growth, the swollen termite population, in harvesting the sparse grass cover, exposed the land surface to more effective processes of lateral movement. Ratcliffe *et al.* (1952) have shown that termites are responsible for causing injury and death to certain trees and hence are effective in decreasing vegetational cover.

A rather less direct action of termites can be seen in those areas where termite mounds control the type of vegetation to a certain extent (Lee and Wood 1971). Thus, in some areas, the growth of vegetation is enhanced on termitaria. This is the case in eastern and southern Africa where a unique type of savannah vegetation is distinguished which consists of discrete islands of woodland growing on large termitaria, surrounded by sparsely timbered grassland. It has also been shown (Thomas 1941) that in certain areas these wooded mounds play an essential role in the vegetational succession that occurs after disturbance by fire, for they act as spreading centres for recolonisation of the inter-mound areas to produce a closed-canopy forest. In contrast, many large mounds, including African and South American examples and particularly those in Australia, do not support vegetation while occupied and, in some cases, even when they are abandoned. Such termite-controlled vegetational variations clearly will have considerable implications for the rate at which processes of lateral movement occur. Australia also provides an example where ants increase the vegetational cover, for Berg (1975) has shown that the spread of about 1500 species of vascular plants is aided by the presence of ant-attracting structures on their seeds and fruit.

In considering the direct action of animals on the processes of lateral surface movement, it is possible to ignore the great majority of animal species that merely inhabit the spaces within the soil material without modifying them in any way. Such an exclusion allows consideration to be restricted to a relatively few groups, i.e. termites, ants, worms, beetles, crustaceans and certain larger animals, particularly those that live in burrows within the soil.

Termites Termites have developed the capacity to burrow and mould structures from soil and organic matter in the process of nest construction to a level unknown in any other group of soil animals and therefore their impact on surface processes must be of considerable pedological importance. Termite mounds are the most obvious evidence of soil reorganisation. They can vary in size from a few centimetres to a maximum of about 9 m in height and 20–30 m in diameter. In terms of the amount of material contained within the mounds, a maximum figure obtained from the southern Congo (Meyer 1960) is 2 400 000 kg/ha, which is equivalent to a layer of soil 20 cm deep over the whole area investigated. The pedological importance of this amount of material is further emphasised when it is realised that 30 per cent of the total surface area is occupied by the mounds. However, such an evaluation gives no idea of the rate at which the soil material is moved. Nye (1954, 1955) showed that, within the forest zone of southern Nigeria, mounds of *Macrotermes bellicosus* are built very rapidly, a mound 60 cm high being built in a month and one 150 cm high within a year. An average mound contains 2500 kg of material and, with an average density of five per hectare, Nye calculated that 1250 kg/ha/year of soil was brought to the surface by termites. Therefore, by this means alone, a layer of soil material 30 cm thick would be deposited on the surface in 12 000 years. Williams (1968), in considering the activity of the termites *Tumulitermes hastilis*, *T. pastinator* and *Nasutitermes triodiae* in northern Australia, calculated that they were responsible for bringing 6000 m^3/ha of soil material to the surface in the course of 12 000 years. Therefore, it can be concluded that, even by considering only mound building by termites, their pedological significance is considerable. Account has also to be taken of those species of termite that transport quantities of soil material above ground for the purpose of packing the eaten-out portions of fallen branches, logs, standing trees and tree stumps, and subterranean nesting species who construct nesting systems entirely beneath the soil surface. In addition, all termites construct systems of galleries and covered runways which they use in searching for food, moisture and soil particles. In the case of one tree-nesting species (*Coptotermes acinaciformis*) in southern Australia, Greaves (1962) has shown that one colony had six main galleries leaving the tree below ground level. In section the galleries were uniformly 3 mm in height and their width varied from 6 to 63 mm, depending on the traffic density; they were generally 15 to 28 cm below the surface, extended outwards up to 30 m from the nest and covered an area of 0·16 ha. Ratcliffe and Greaves (1940) studied the subterranean gallery system associated with *Coptotermes lacteus* and *Nasutitermes exitiosus* in southeastern Australia. In the case of the first species two mounds had nine and thirty-six main galleries radiating from them; five mounds associated with the second species had from eighteen to thirty-six galleries associated with each of them. In general size and range the galleries were similar to those described for *C. acinaciformis*. However, there have been no quantitative estimates of all termite galleries in a given area or of the amount of space they occupy.

In the case of the subterranean nesting species, little is known about their gallery systems, but there is no reason to think that they are not extensive. Again, there is no quantitative information on the amount of soil material transported above ground either for constructing covered runways or for packing inside branches and logs. However, it is probable that in areas where many species of termite are exploiting a wide range of food resources (humus, plant litter, grass and herbs or dead and living wood), the amount of soil material being affected is likely to be as great as that involved in the mounds of mound-building species.

In the process of moving this considerable amount of soil material, termites appear to show some degree of selectivity of soil particles, depending upon their size. In general it would seem that, if any selectivity occurs at all, clay is preferentially accumulated in sand-rich materials and *vice versa*. Care has to be taken in assessing the degree to which this process has occurred, particularly in those cases where the soil consists of a sand-rich surface over a more clay-rich subsoil. In many such instances the mound has the same particle size distribution as that of the subsoil, i.e. the subsoil as a whole has been moved without any selectivity (Lee and Wood 1971).

However, there are certain cases in which differential accumulation of clay is undoubted. Nests of *Macrotermes natalensis* within the Kalahari sand contain from 8 to 13 per cent clay, whereas the clay content of the sand down to a depth of 84 m varies from 1·8 to 5·4 per cent (Bouillon 1970).

A rather different example of clay enrichment is provided by nests of *Drepanotermes rubriceps* in South Australia which have more clay and less sand than the surrounding surface layer of soil material. Such evidence suggests that at least some of this material was derived from the more clay-rich subsoil. Identification of clay minerals, however, showed that the clay of the nests was the same as that of the surface soil and differed from that of the subsoil. The termites must therefore have altered the ratio of clay to coarser materials by removing some of the sand to the surface and building with the residue (Lee and Wood (1971)).

Stoops (1964) has shown that particle sorting by termites may become rather more complex. In mounds of *Cubitermes* spp., compared with neighbouring soils, there was an enrichment in particles of < 100 μm and > 500 μm and a considerable improverishment of particles of intermediate size. Stoops showed that the termites swallow fine soil particles, carry them in the crop and then excrete them, while larger particles are carried in the mandibles. It was concluded that particles in the 100–500 μm range are too large to be swallowed and too small to be conveniently carried.

In the process of moving soil material around and mixing it with organic matter in various forms in order to build nests, runways and galleries, termites are responsible for the development of new fabrics. It is possible to differentiate between those fabrics that are mostly inorganic and those dominantly organic. The inorganic fabrics make up the greater part of the outer walls of mounds and the infilling of former galleries and two main types have been recognised as being formed by a range of termites in both Africa and Australia (Lee and Wood 1971). In the first the workers carry a grain of sand in their mandibles and finer particles (clay to fine sand) in the crop. On reaching the site of construction the sand grain is placed in position and the contents of the crop moistened with saliva, used as a cement around it. Initially, the new construction has an open spongy appearance with many tunnels and pores. Subsequently, this open fabric is filled in with more soil particles until it becomes dense and massive. Another method by which a dense and massive fabric could develop has been observed by Greaves (1962) in the formation of subterranean galleries by *Coptotermes* spp. by compression of soil away from a central point to form a compact layer around the galleries. The second type of fabric is one in which lenticular parcels of soil material (0·5–2·0 mm long and 0·05–0·5 mm wide) with a thin (0·02 mm) coating of dark amorphous excreta are stacked in a sub-parallel arrangement. The development of this fabric has been observed under laboratory conditions. The workers put an individual lenticle of soil, moistened with saliva, in place and then excrete a drop of thick fluid over it. A variant of this fabric occurs in the material forming the nursery walls of mounds; the lenticles of material in this case are moulded from masticated wood.

The fabrics that are dominated by organic material are particularly well developed in nursery areas and gallery linings and again two main types can be distinguished. In the first the material consists of closely adpressed faecal pellets ranging in size from 0·5 to 1 mm. Within the pellets, comminuted fragments of wood are visible. The second fabric type consists of thin parallel laminae, apparently laid down in the fluid state and probably consisting of excreta. There is some variability in this material; in some cases it is possible to recognise many very finely comminuted wood fragments which have apparently passed through the termites' alimentary system without being digested; in other cases there is a small but constant proportion of mineral grains which must have been ingested by the termites.

Termites, then, are responsible for a considerable reorganisation of the soil mantle and must inevitably have a considerable impact on the processes of lateral surface movement. The kind of impact involved is best illustrated by two particular examples. Glover *et al.* (1964), during the course of an ecological survey of the Loita Plains area of south-western Kenya, investigated a vegetation pattern visible on aerial photographs. The elements of the pattern varied from a circular shape, 6·5–10 m in diameter on the interfluves, to ovate or elongate ellipsoids, 6·5–10 m broad and up to 14 m long on slopes down to the drainage lines. This arrangement gives a 'peacock feather-like' arrangement (Fig. 8.1). Each of these elements is centred on a termite mound of *Odontotermes* sp. up to 0·6 m high and there are about 650 mounds per square kilometre. On the centre of each mound there is a growth of low shrubs. Peripheral to the mound, in the case of the circular elements,

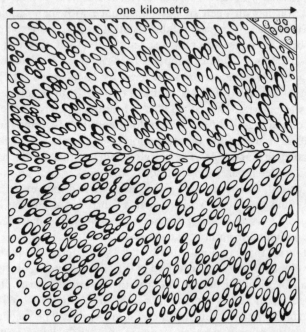

one kilometre

Figure 8.1 Distribution of termitaria on the Loita Plains of South-west Kenya (after Glover *et al.* 1964, by permission of Blackwell Scientific Publications).

or on the upslope margin of the ellipsoids, is a zone of tall grass. The downslope pointing long tail of the ellipsoids has a short, sparse grass cover (Fig. 8.2). The extremely poor vegetational cover in this tail zone is because of the downslope wash of material from the termite mounds, which must be readily dispersible for it has formed a hard, compact surface layer. With such a concentration of mounds all contributing to such an effect, the alteration of run-off characteristics of the region as a result of termite activity must be assessed as being very considerable.

The production of the hard, dense, massive fabric is another aspect of termite activity that has a profound effect on surface processes. This aspect has been stressed by Watson and Gay (1970) in their work on the role of grass-eating termites in southwestern Queensland, referred to previously. The nests of *Drepanotermes perniger* have hard, dense, massive caps, shaped like shallow inverted saucers, 2–3 m in diameter, lying immediately below the shallow, friable topsoil and covering approximately 20 per cent of the total ground area. The harvesting activity of the termites bares the ground surface and leads to the erosion of the surface by running water until the hardened surface of the nest is exposed. At this point run-off occurs as from a solid rock surface and no seeds can find a spot to lodge and germinate. In this state the hard, dense nest material can persist possibly up to

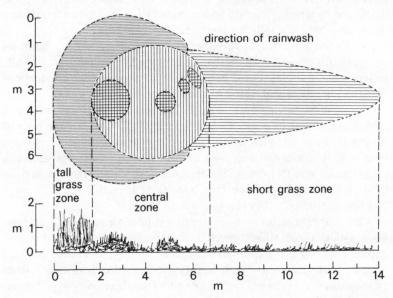

Figure 8.2 Vegetation, soils and topography of single termitaria (after Glover *et al.* 1964, by permission of Blackwell Scientific Publications).

100 years without breakdown. Thus in this case the development of a new soil fabric as a result of termite activity has had a profound effect on surface processes.

From the discussion above it is apparent that wherever termites occur they have very considerable pedological significance. The great majority of them live in tropical and sub-tropical regions with very few extending to the extremes of their range at about 45° north and south of the equator.

Ants In contrast to termites, ants have a very extensive distribution, from the arctic regions to the tropics, from the tree line down to sea level and from the most humid to the driest of situations. In addition, ants are not necessarily so confined to such a subterranean existence within the soil as are the termites. Even so, there are sufficient ants burrowing into the soils of most of the land areas of the world for them to have considerable pedological importance.

Insofar as ant nests occur in soils, they always consist of a subterranean portion comprising a number of more or less irregular excavations. In most cases these excavations are fairly widely separated, but in some cases they are fairly concentrated. Such subterranean developments may or may not be associated with a superstructure above ground level. It seems probable, therefore, that fabrics, similar to those developed by termites, should be associated with ants. However, to date there have been virtually no investigations of this idea.

Ant excavations consist of either chambers or galleries. Chambers are generally larger with flattened floors and vaulted roofs, while galleries are more tenuous, being more or less tubular connections between chambers and nest openings. Chambers vary greatly in size; those of the harvesting ants, *Messor barbarus* and *arenarius*, in North Africa are 15 cm in diameter and 1·5 cm high, while in the case of fungus-growing ants such as *Atta texana* they are sometimes 50–100 cm long and 30 cm broad and high (Wheeler 1910). Galleries generally range from 2 mm to 1 cm in diameter, but occasionally they reach 6 cm in size. The depth to which ants operate is generally not more than 2 m but occasionally they reach to 5 m or more.

Where superstructures are developed it is possible to differentiate two main types: those which develop a crater-like form around a nest entrance and those which have a much more solid and continuous mound. In the case of the craters, they are formed only of debris from excavations within the soil and can be either single grain or agglutinated into pellets. The mounds are formed to a much greater extent of material collected from the surrounding surface area and within the mounds a number of chambers and interconnecting galleries are generally found.

Attempts have been made to assess the impact of ants on soil material by investigation of the rate of mound development and decay. For instance, Waloff and Blackith (1962) examined the development of mounds of *Lasius flavus* in Berkshire, England, over a period of several years and concluded that an area equivalent to the total surface area would be occupied in a period of about 100 years. Thorp (1949), reporting on ant activity throughout the sub-humid and semi-arid areas of the USA, made a conservative estimate of the amount of soil material contained in ant mounds at any one time as being 3400 pounds (1550 kg) of soil per acre. Baxter and Hole (1967) attempted to be more specific about the dynamics of a similar situation by considering the activities of a particular species of ant (*Formica cinerea montana*) on a particular prairie soil (Tama silt loam) in southwestern Wisconsin. They reported 1531 mounds/hectare with an average diameter of 37 cm; height 15·3 cm and volume 0·02 m³. The mounds are calculated to contain 34 m³ of material per hectare and cover 1·7 per cent of the ground surface. Taking the average occupancy time of mounds as twelve years, an area equivalent to the whole of the surface would have been occupied by mounds in about 700 years. From these examples it can be concluded that such activity must have considerable pedological importance. However, as mentioned previously, ants do not invariably produce mounds, so that such an estimate can only be taken as providing a minimum figure of possible soil disturbance. This is so even if only a single species is considered, for Waloff and Blackith (*op. cit.*) showed that within the area they investigated there were patches of lighter-textured, better-drained soils in which the nests of *Lasius flavus* were entirely subterranean. This ant does not build mounds in southern Europe or eastern North America, but does so in the Mississippi Valley and the Rocky Mountains (Wheeler 1910). More importantly, however, it can be said that by far the greatest number of ants' nests are excavated in the soil under stones and logs, etc. without any recognisable superstructure being produced. In addition, even in the case of ants which usually construct mounds, they also are entirely subterranean in the early stage of colony development. In this incipient stage ants take the greatest care to conceal the situation of their nests. The excavated soil pellets are therefore carried some distance from the nest opening and scattered irregularly over the surface. Thus calculating the amount of soil involved is difficult, if not impossible.

It can be concluded then that, despite the lack of a great deal of detailed knowledge of ant–soil interactions, the amount of soil disturbance and rearrangement that can be ascribed to ants is at least as great as that caused by termites and undoubtedly is spread over a wider geographical area.

Earthworms   There has been a great amount of work on earthworms, but for the most part this has concentrated on either their biology or the implications of their activity in terms of soil fertility. Data on earthworm–soil interactions are generally incidental to these other investigations and therefore information in this area is relatively sparse. It is unfortunate that the lead provided by Darwin (1881) has not really been capitalised upon.

There are about 1800 species of earthworm with an almost worldwide distribution. They vary in size from a fraction of a centimetre to exceptional individuals of *Megascolides australis*, up to 2·75 m in length and 3 cm in diameter. The most common earthworms in Europe, Western Asia and most of North America belong to the family Lumbricidae. Other regions of the world are dominated by different families: thus the megascolecid group is the most important family in southern and eastern Asia and Australasia. However, a considerable number of lumbricids have been introduced into India, Ceylon, South Africa, Australia and New Zealand as a result of European colonisation. In this situation these earthworms have often multiplied rapidly and supplanted the native species. Because of this dramatic human intervention, it is difficult to assess the natural impact of earthworms on soil material.

Earthworms feed on dead or decaying plant remains, including both leaf litter and dead roots. In the consumption of this plant debris most earthworms consume soil material, for reasons that are not understood. In addition, certain species consume a great deal of soil material in the process of making burrows. The soil material ingested is voided either at the surface in the form of casts or within the body of the soil, often within burrows. Permanent burrows are restricted largely to those species that penetrate deep into the soil and many of these only do so when the food supply at the surface is no longer adequate (Evans 1948). Such burrows have smooth walls cemented together with mucous secretions and excreted soil pressed into voids within the soil material. They range in diameter from 3 to 12 mm, but it is not certain whether earthworms increase the size of their burrows as they grow, or make new ones. Satchell (1967) showed that, under old pastures at Rothamsted, *Lumbricus terrestris* formed burrows to a depth of just over 1 m, *Octolasium cyaneum* to about 55 cm, and *Allolobophora nocturna* to 33 cm. These burrows are, for the most part, vertical, branching extensively only near the surface. All other species of earthworm, which are usually of a smaller size, operate within the surface 20 cm and do not form permanent burrows.

It is usually the burrowing species that produce casts on the soil surface near the exits of their burrows. Of the eight to ten common field species of lumbricids in England, only three – *Lumbricus terrestris*, *Allolobophora nocturna* and *A. longo* – particularly the latter two species, produce casts on the surface of the soil. Usually they cast more on heavy soils than on light, open ones, because in the latter case much of the excreted material is packed within the voids of the soil. Worm casts can vary a great deal in form but are often characteristic of a particular species. They range from the small, heterogeneous casts of *A. longa* and the individual pellets of *Pheretima posthuma* to the red clay chimneys, 10–12 cm high and 4 cm in diameter, of the African worm *Dichogaster jaculatrix* (Edwards and Lofty 1972).

In terms of evaluating earthworm–soil interactions the most important criterion is the rate at which soil material is ingested. As indicated previously, one of the greatest difficulties to be faced in assessing this amount is that not all earthworms produce surface casts and it is difficult to recognise and take account of soil material voided by earthworms within the body of the soil. One of the first attempts to solve this problem was made by Evans (1948). He demonstrated that worm casts were formed for the most part by *A. longa* and *A. nocturna*. Evans assumed that all the soil ingested by the surface-casting species was ejected on the surface and that, weight for weight, other non-surface-casting species consumed an equal amount of soil but voided it within the body of the soil. Knowing the total weight of the worm population it was then a simple matter to calculate the weight of soil material ingested by the non-surface-casting species. On this basis it was shown that the weight of surface earthworm casts produced per annum varied form 250 g/m^2 to 6250 g/m^2 and the weight of soil voided by earthworms beneath the surface varied from 500 g/m^2 to 5250 g/m^2.

Evans realised that these figures must be taken as minimum values since the amount of soil material which occurs as surface casts does not represent the total amount of soil consumed by surface-casting species since some of the material ingested is used to line the burrows which they construct in the subsoil. Also, during frosty spells in winter, surface cast production is very meagre or non-existent because the worms have retreated into the lower layers of the soil and presumably void ingested material there. In addition, Satchell (1967) pointed out that the potassium per manganate method used by Evans in determining the total population of earthworms is limited by the penetration of the solution and the variable response of the earthworms to it. The maximum figure obtained by Evans for soil material ingested by earthworms was from an old permanent pasture where surface casts accounted for 6150 $g/m^2/$annum and subsoil voiding for 2925 $g/m^2/$annum. Taking the combined figure of 9075 g, and considering that a square metre of soil to a depth of 10 cm would have a weight of 170 000 g (accepting a soil density of 1·7), it can be calculated that the time required for this mass of soil material to pass through the alimentary tracts of the earthworm population would be 18 to 19 years. From previous statements this can in no way be regarded as a maximum figure; therefore, it must be concluded that the impact of earthworms on soil material is very considerable.

Satchell (1967), working on the earthworm population in some English Lake District woodlands, has provided the means of solving this problem in a rather different way. It was shown that, on average, individuals of *Lumbricus terrestris* contained 100 milligrammes of soil per gramme of body weight. Food passes through the gut of this earthworm in 20 to 24 hours, so that the rate of ingestion of soil material would be 100–120 mg/g/day. Such a rate varies with soil temperature and internal factors affecting the metabolic rate, but if it were maintained for 200 days in the year, the total annual soil intake by a population which had a mean biomass of 120 g/m^2 would be about 2640 g/m^2. On the same basis as previously (i.e. taking soil to have a density of 1·7 and considering the surface 10 cm) the time required for this mass of soil material to be ingested by earthworms would be 64 years. Even though this figure is three times that resulting from Evans' work at Rothamsted, in terms of the effect on soil material it must still be regarded as being of the utmost significance.

If the data for earthworm–soil interaction in temperate regions are limited, those from tropical regions are almost non-existent. Two sets of data, both from the Ibadan area of Nigeria, Nye (1955) and Madge (1969) appear to be among the most reliable. Nye worked in a forested area in which the dominant species of earthworm is *Hippopera nigeriae*. It was calculated that this earthworm brought to the surface 5100 $g/m^2/$ annum of soil material. For a surface layer of 10 cm this means a throughput time of 34 years. However, Nye was concerned only with the addition of material to the surface and therefore there is no attempt to evaluate the possible subsurface voiding of soil material by *H. nigeriae* or by any other species. Madge confined himself to investigating the activities of earthworms in grassland areas and not under the original forest. Two main species were found – *Hyperiodrilus africanus* and *Eudrilus eugeniae* – the first confined to soil in shaded grassland and the second to open grassland areas. In both cases it was estimated that, with a population of thirty worms per square metre, each one of which voided 3·2 g of surface casts every day for the six months of the wet season, something of the order of 17 500 g of material per square metre were voided at the surface in one year. For the surface 10 cm this means a throughput time of just under 10 years. In assessing this very high turnover rate compared with that of Nye, account must be taken of the fact that this is a grassland area within the West African forest zone, a highly artificial situation. Another point made by Madge is that, although the mature forms of both *H. africanus* and *E. eugeniae* are small compared with the surface casting of European species, the casts produced by these Nigerian species are considerably more impressive than those voided by earthworms in temperate regions. Again, as in the case of Nye's work, no attempt was made to evaluate the effects of the subsurface activity of earthworms, although mention was made of deep subsoil activity during the dry season.

We now turn to the question of the possible sorting of soil material by earthworms. As a result of mechanical analyses of worm casts, Stockli showed that particles greater than 2 mm in diameter were not ingested by earthworms (Satchell 1958). It seems reasonable, therefore, that in soil material strongly influenced by earthworms particles coarser than 2 mm would be less abundant compared with underlying material which has not been so strongly affected. Evans (1948) demonstrated such an effect to a depth of 10 cm in soil material that had been under pasture for 300 years at Rothamsted, but it is impossible in these circumstances to eliminate the possibility that other processes, such as generalised downslope movement, may not also be involved in such differentiation. Nye (1955) also mechanically analysed worm casts and showed that they contained no particles greater than 0·5 mm in diameter and only a low proportion with diameters between 0·2 and 0·5 mm; the surface horizon of about 3 cm thickness was ascribed to such worm activity for it was identical in terms of mechanical analysis and contrasted with the much coarser-grained underlying material.

It has been suggested by several workers (Satchell 1958) that earthworms bring about a comminution of the soil due to its continual passage through their alimentary tracts. If this is so, the reduction of the amount of coarse sand in material that has been voided by earthworms may be due in part to this cause. However, considerable doubt has been expressed about the evidence for such intestinal grinding.

In the process of ingesting and voiding soil material, earthworms must necessarily have a considerable effect on soil fabric. Apart from the work of Barratt (1964) and of Jongerius and Schelling (1960), most reports are given indirectly in terms of aggregate stability and porosity rather than in terms of the actual fabric produced. From the evidence of wet-sieving and water-drop stability tests, it has been generally agreed that worm casts contain more water-stable aggregates than soil material not in this form. There is a variety of explanations of how such water-stable aggregates are formed, but the most generally accepted one is that the stabilising materials originate from the microbial populations which develop either within the material in the gut or subsequently in the casts. The work of Parle (1963), which showed that changes in the stability of worm casts can be related closely with the development of fungal hyphae, seems to indicate that this is the principal factor involved, rather than bacterial gums, as suggested by Swaby (1950). However, in terms of soil material as a whole, the effect of this increased stability must be of minor importance for worm casts retain their stability for a period of a month or so at the most; taking an average turnover time for the surface 10 cm of soil as being 60 years, it follows that well under 1 per cent of the soil is in this stabilised form at any one time.

Earthworms are considered by some to have a considerable effect on soil aeration. However, species that void material below the soil surface can only open the soil at one point at the expense of consolidating it at another or by filling in existing air spaces; the only activity that can increase soil air space is surface casting. From Evans' (1948) figures, the volume of soil ejected annually onto the surface by earthworms is only 0·5–6·0 per cent of soil in the top 10 cm with a total pore space of 40–59 per cent of its volume. Since the rate of casting on the Rothamsted pasture appears to be average for western Europe, it seems the effect of earthworm soil aeration is of only minor importance (Satchell 1967).

Drainage effects, however, are probably much more considerable, for there is a great deal of evidence such as that of Slater and Hopp (1947) who showed that infiltration is generally much faster in field soils with high earthworm populations than where populations are low. This field evidence is supported by the result of laboratory experiments such as those of Guild (1955), in which it was shown that water passed through containers of light sandy soil in two days when the soil had been affected by worm activity and in eight days when worms were absent.

Crustaceans To anyone who is familiar with soil materials in coastal areas, the way in which crabs move material around is undoubted, but to date such conclusions are based on casual and generally very sporadic observations. Thorp (1949) has provided rather more data in the case of crayfish, which are active in poorly-drained soils from the Gulf Coast north to the Canadian border. In southern Indiana crayfish build chimneys on the soil surface, with deep subsoil material. It was estimated that 800 g/m^2/annum were being added to the surface by this process. However, observations were not extended sufficiently in time for the figures to be anything more than first approximations.

Mammals Burrowing mammals disturb soil material to a considerable extent. This is particularly noticeable in areas of sparse vegetational cover, but it is probable that the impact is just as great in forested areas. Thorp (1949) dealt with the effect of prairie dogs and badgers on soil material in Colorado. The soil is developed on loess 1·8–3 m thick with a silt–loam texture overlying a gravelly-sand layer. All the burrows penetrate down to the gravel layer, for the surface mounds are dominantly sand and gravel. The large mounds are up to 7 m in diameter and 45 cm high. Overall some 7·6–10 kg/m^2 of mixed sand and gravel have been brought to the surface from a depth of about 3 m. Thorp also observed at Lincoln, Nebraska, the effect of pocket gophers and ground squirrels, which from time to time appear in large colonies. It was calculated that they were responsible for adding to the surface seven kilogrammes of soil material per square metre, in the form of mounds which individually varied in weight from 7·5 to 150 kg. In both these cases, however, no indication is given of the rate at which these additions took place.

Price (1971), working in the Yukon, has attempted to evaluate the effect on near-surface processes of the arctic ground squirrel in an alpine environment. The effect is markedly concentrated on south-east facing slopes and even on these slopes it is restricted almost entirely to protected sites in front of solifluction lobes. Within such a restricted area the squirrels have added two kilogrammes of soil materials per square metre of surface per annum. However, it could well be that a greater effect on surface processes stems from the burrows not only of the squirrels but also of the abundant lemmings, mice and shrews; these burrows provide channel-ways for meltwaters which are responsible for a considerable amount of subsoil sapping.

A much more comprehensive attack on this problem is that by Imeson (1976). He considers the effect of burrowing moles and voles in a forested area of the Luxemborg Ardennes, as part of a comprehensive investigation of slope processes. The rate at which material is being raised to the surface was determined as being 1940 g/m^2/annum, which is remarkably similar to the figures obtained by Price (see above). However, the areally restricted nature of Price's sites compared with those of Imeson needs to be emphasised. It would seem that in this

particular example the activity of these small mammals, if taken together with that of earthworms, is responsible for providing most of the
material involved in subsequent downslope transport. Imeson stresses
the preliminary nature of his results, but even so it is reasonable to
conclude that the burrowing activity of animals is one of the major
slope processes occurring under humid deciduous woodlands today
and, as such, warrants more attention than it has hitherto received.

In addition, Imeson draws attention to the possible importance of
rooting by wild pigs which are capable of moving a considerable mass
of soil material within a few days and leaving behind a characteristic
microtopography of mounds and pits 40 cm or so deep. He saw similar
mounds and hollows, often partly litter covered or overgrown, which
suggested similar rooting activity during the preceding years.

The author has observed the impact of wild pigs on slope processes
under the high tropical forests of Sabah, Malaysia. On the east coast
there are few natural predators of these animals, so that they move
about in large herds. They naturally like to wallow in mud and scrape
the mud off afterwards when it is dry. The soil material in this area is
clay-rich and under the intense rainfall there is a considerable movement
of material downslope in the form of small slips, despite the dense
vegetational cover. The head of each such failure is marked by an
arcuate fracture which leaves an upslope escarpment up to 60 cm high.
Within the arcuate fracture there is slight ponding of water on a clay
subsoil – an ideal situation to be exploited by the pigs which create a
wallow in the swampy area and use the slip scarp to scrape off the mud.
Continued use extends the wallow for the activity of the pigs, destroys
the soil fabric and more and more water is ponded within the wallow.
Finally, it exceeds the stability of the slope, perhaps due to a particularly heavy downpour of rain; the downslope lip of the wallow is
breached and the resulting mudflow drains the wallow. The drained
wallow is then an ideal site for subsequent gully initiation. In view of
the concentration of pigs in this area, their erosive potential is possibly
very considerable but, again, quantitative data are lacking.

Despite this general paucity of accurate data, it is apparent that the
direct impact of the larger animals on soil material can, locally at least,
be of considerable significance. The case becomes even stronger when
it is realised that to date only very few of the burrowing animals have
been considered in this light. Thus, within Australia, the wombat must
have a considerable effect in certain areas and from Melton's work in
Africa (1976) the same conclusion can be reached concerning the
aardvark.

This chapter has demonstrated that biospheric interaction with soil material must have considerable pedogenic significance. In Chapter 6 it was shown that processes of lateral surface movement are also of the greatest significance. Because of the lack of quantitative data concerning the absolute rate at which both these sets of processes operate there is at present no way of evaluating their joint impact on soil material.

9

The factors of soil formation

In previous chapters the processes of soil formation have been discussed. The relationship between these processes and the factors of soil formation will now be considered. There are generally acknowledged to be five such factors: lithospheric material, topography, the biosphere, climate and time (to avoid the definitional tangle associated with the term **parent material,** the more generalised **lithospheric material** is used). The establishment of this relationship is of the utmost importance for it will integrate soil formation more generally with the environment and enable a world view to be taken. The necessity of linking processes and factors has frequently been ignored for, as Crocker (1952) stated, studies in soil genesis are, for the most part, concerned either with soil dependence upon environmental factors or with the actual processes involved in the formation of soil material, but rarely with both at the same time. Crocker's paper itself is a good example of the first approach and the first eight chapters of this book of the second.

Recently, Birkeland (1974) has dealt with both processes and factors of soil formation and stressed the need to distinguish clearly between them. The distinction as he sees it is that the processes form the soil, while the factors define the state of the soil system. This suggestion is now examined in greater detail by considering the relationship between the processes of soil formation and each of the factors in turn. It is possible to introduce some rationality into the order in which the factors are discussed by adopting Joffe's (1936) suggestion and dividing them into two groups: (a) those which provide mass only, together with those that control the position of the mass (the passive factors) and (b) those that supply the energy (the active factors). The first group, comprising lithospheric material and topography, will be discussed before the second group which consists of climate and the biosphere. This leaves until last discussion of the fifth factor, time.

Lithospheric material

In considering the material making up the lithosphere as a factor of soil formation there is apparently an almost infinite number of starting points, judging from the great number of rock types that have been recognised. However, as was shown in Chapter 1, it is possible to rationalise this complex problem considerably when it is realised that everything is initially derived from igneous rocks. In terms of extent of outcrop of igneous rocks on the land surface of the world there are two types which are overwhelmingly dominant: granite and basalt. The occurrence of one or other of these rocks controls, to a considerable extent, the way in which the processes of soil formation operate.

Granite, when subject to epimorphism, gives rise ultimately to a material which is a mixture of quartz-sand and newly-formed clay-sized particles. When such material is affected by processes of lateral movement in which water is the dominant factor, there is a marked sorting of the coarse and fine particles into distinct bodies of material. In contrast, when basalt is subject to epimorphism, as there are no minerals strongly resistant to weathering, only fine-grained, newly-formed particles of clay minerals and sesquioxides develop from the feldspars and ferro-magnesian minerals. The resulting uniform, fine-grained material is markedly different from that derived from granite and these differences become even more marked when processes of lateral surface movement occur, for in this case there is no possibility of differential sorting, no matter how the material is moved.

Granites and basalts, however, occupy something less than 30 per cent of the land surface of the earth. The remainder is underlain by sedimentary rocks of which sandstones, shales and limestones are the most important types. When such rocks are subject to epimorphism the reactions tend to be superficial and are seldom of a fundamental nature. Sandstones require only the cement to be removed for a soil material to be produced, while shales, if composed of kaolinites and montmorillonites, require only water penetration between the platelets. Even in the case of fine-grained sediments in which the minerals have been diagenetically or metamorphically altered, as long as the sheet structure is preserved, only a relatively superficial type of alteration such as that described on page 36 is required. Limestone is somewhat different in that calcium carbonate is unique among commonly-occurring minerals in being so readily soluble in water. However, the soil material associated with limestone is merely the insoluble residue from this process; the clay minerals and the sesquioxides that occur are, to a large extent, inherited, with little formation of new minerals.

It is possible to describe the epimorphism operating on granite and basalt as *primary*, for what is involved is the formation of new minerals that are fundamentally different in structure from the original minerals. In contrast, the epimorphism of sedimentary rocks can be regarded as *secondary*, for new mineral formation, if it occurs at all, tends to be of minor importance. In the case of metamorphic rocks, the higher the grade of metamorphism to which a rock has been subjected, the more will the epimorphism required to produce soil material from it be primary rather than secondary.

Another factor involved in the changes of sedimentary rocks to soil material is **inheritance,** for the dominance of superficial change means that a considerable number of features, both as regards mineralogy and fabric, are preserved in the soil material. There is a considerable increase in the inheritance factor when more unconsolidated materials such as sand and clay deposits are considered. In an alluvial deposit it is often impossible to differentiate between what is inherited and what is due to current processes, for at this point the two merge.

The selection of granite and basalt as central examples of the reaction of rocks to surface processes, despite their relatively restricted outcrop, is justified by consideration of the way in which sedimentary rocks react. Sandstones and shales are often interbedded and both outcrop on many hill slopes. The result of secondary epimorphism and lateral processes in such a situation is very similar to the effect of the same processes on granite, the only difference being that it is achieved more quickly. In the case of soil material derived from limestone, the inherited material is generally of a fine-grained nature and is therefore in many ways similar to that produced by primary epimorphism of basalt.

From this discussion it can be concluded that the nature of lithospheric matrial controls, to a considerable extent, the way in which surface processes will operate.

Topography

Considering topography in the most general way, as a factor controlling the processes of soil formation, it can be said that the dominant topography on the surface of the earth is associated with river valleys, which vary in a systematic manner from drainage divide to streamline. Within this context it is possible to recognise two extremes of topographic variation which are responsible for a marked difference in the operation of, and the balance achieved between, soil-forming processes. The first case consists of relatively gentle topography and the second of strong relief with a dominance of extremely steep slopes. Material moving across these slopes from water divide to stream channel requires a relatively long period of time in the first case and a short time period in the second. Such a difference has considerable implications with regard to both soil processes and the soil material that results.

In the case of gentle topography, the long residence time of soil material means that epimorphic processes can proceed a considerable way towards their end points. For example, the only primary mineral remaining from the weathering of granite would be the highly resistant quartz, while new mineral formation would be at a maximum. As discussed on page 44, the overall result would be a material of very contrasted particle size: sand-sized residual quartz and clay-sized newly formed minerals. The lateral processes of movement involved in such gentle topography are, for the most part, those in which water is dominant (p. 54). Such processes, acting on the material of contrasted particle size resulting from epimorphism, will produce well-sorted bodies of material.

The balances achieved between epimorphism and lateral processes within a landscape of gentle topography give rise to three types of site (Fig. 9.1) with markedly different soil material. At site (1) epimorphic processes are associated with little or no lateral movement and the resulting soil material can be regarded as RESIDUAL. At site (2) there is much more of a balance between epimorphism, which continues to act on the bedrock, and the processes of lateral movement, which are dominant in the near-surface layer. Typical soil material from this type of site is described on page 62. As the mobile top soil is such a characteristic of this site, it would be appropriate to refer to it as TRANS-PORTATIONAL. At site (3) the processes of lateral movement with concomitant sorting are dominant, while epimorphic processes are of minor importance at most. Such sites are DEPOSITIONAL and materials typical of them have been discussed in Chapter 6 on page 56.

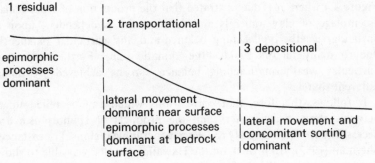

Figure 9.1 Topographic sites.

Circumstances are markedly different on steep topography, for the short residence time of soil material means that the epimorphic processes can proceed only a short distance towards their end points. After weathering, a great number of primary minerals in addition to quartz will remain, while new mineral formation is minimal. The overall result will be material of varied mineralogy and particle size. The lateral processes of surface movement involved with such steep topography are those in which mass movement is dominant. Such processes acting on material produced by minimal epimorphism result in a mass of unsorted material of varied mineralogy. In terms of the material that occurs on such steep slopes the absolute dominance of lateral processes of mass movement means that the upper slopes often consist of bare rock surfaces, while the lower slopes are covered in a mass of unsorted debris with little sign of *in situ* alteration caused by epimorphism.

Climate

Climate as a factor controlling the processes of soil formation is generally considered in terms of precipitation and temperature. There is no doubt that epimorphism is fundamentally controlled by these two parameters for, if H_2O were not present in the form of water at the surface of the earth, weathering, leaching and new mineral formation, as previously defined, could not occur. However, difficulties are encountered when attempts are made to go beyond such a basic position and develop connections between particular climatic indices (such as various forms of precipitation/evaporation ratio) and the products of epimorphism over large areas of the Earth's surface. Such correlations presuppose that there is a direct and unique relationship between climatic indices and the type and quantity of the epimorphic products. That this is not necessarily the case can be seen from an examination of Figure 4.4 where it is demonstrated that the production of a particular assemblage of clay minerals and sesquioxides is dependent upon a particular weathering/leaching balance and any particular balance is due to many factors besides the climatic ones. Furthermore, any particular weathering/leaching balance can be achieved by many different routes.

It follows, therefore, that correlations which can be relied upon between climatic parameters and the products of epimorphism are necessarily confined to fairly restricted geographic regions. For instance, Sherman (1952), from work on the Hawaiian basalts, was able to show that montmorillonite is the dominant clay mineral produced when precipitation is less than 100 cm/annum, kaolinite between 100 and 200 cm, and free sesquioxides above 200 cm precipitation. More recently, Hay and Jones (1972), working on a particular basaltic ash in Hawaii, were able to show that in terms of weathering only the volcanic glass is altered when the rainfall is less than 115 cm/annum; between 115 and 255 cm plagioclases show signs of alteration while, with a rainfall in excess of 255 cm/annum, all minerals are altered to some degree. In terms of new mineral formation it was demonstrated that montmorillonite occurs in the 25–65 cm rainfall zone, while gibbsite occurs in areas receiving more than 370 cm of rainfall per annum.

Considering processes of lateral movement, a rather more obvious relationship can be seen between rainfall and the processes involved. The more rainfall is evenly distributed and of moderate intensity, the more will processes of lateral movement be those in which water is the the dominant factor, as outlined on page 55. However, the more that rainfall occurs in short periods of great intensity the more will mass movement in the form of slides and mudflows become common (p. 59).

At low temperatures where ice is the common form of H_2O, epimorphism is restricted to those areas where meltwater is available, for the processes of weathering, leaching and new mineral formation are absolutely dependent on the presence of H_2O in the form of water. In terms of the lateral surface movement and deposition of materials by glaciers and ice sheets, no matter whether this material is derived by frost wedging, bedrock grinding or from former unconsolidated surface materials, there is no possibility of sorting during transportation (unless subglacial meltwater streams are present) and the depositional moraines are very like the materials resulting from mass movement. In periglacial regions, where there is a seasonal dominance of water in near-surface layers, mass movement is the dominant lateral process. It is possible to conclude, therefore, that while the overall importance of climate as a factor of soil formation cannot be disputed, the consequences of its differential impact from one part of the Earth's surface to another are indirect and rather difficult to discern.

The biosphere

The importance of the biosphere in the processes of soil formation has been amply demonstrated in Chapters 7 and 8. However, the status of the biosphere as a factor of soil formation remains difficult to evaluate, for it has been established previously that the biosphere is involved in a great number of complex interactions with the material of the lithosphere, in which feedback relationships are so common as to make definition of the role of the biosphere as a factor almost impossible. Problems are also encountered because the biosphere is so dependent on climate, which again makes the disentanglement of a factorial biospheric effect very difficult. However, while keeping these difficulties in mind, it is possible, by considering certain aspects of geographic distribution, to recognise that vegetation at times acts in a factorial manner.

First of all, considering the broadest aspect of vegetation distribution, there is an immense contrast between vegetated and unvegetated areas. Within vegetated areas any of the processes described in Chapter 7 could occur; in the absence of vegetation they would be impossible. There is a similar contrast when lateral processes are considered. Within vegetated areas the processes of detachment, transport and deposition are slowed down compared with non-vegetated areas. The increased stability of soil fabric, which is general within vegetated areas, further increases this contrast.

In a rather more geographically restricted sense there is a considerable contrast in the balance between the processes of soil formation beneath grassland and forest. The fine, uniform root mesh which penetrates to a considerable depth beneath grasslands means that the soil fabric is much more stable and the fabric elements themselves are held together to a much greater extent than is the case under forests where roots are much more concentrated in the near-surface zone. This tends to make mass movement more common under forest than grassland and this tendency is reinforced by the effect that the vegetation has on precipitation. Dense uniform grass cover close to the ground causes water to reach the surface in a uniform and gentle manner, while in forests the interception of rainfall by leaves some distance above the ground means that water from leaf drip will impinge upon it with considerable momentum. Trunks also tend to concentrate downward flow around the base of the trees. In addition, the great amount of soil disturbance associated with tree-fall has no counterpart in grasslands.

On a much smaller scale, at the level of a particular vegetation community or even that of a single plant, the vague boundary between vegetation acting as a factor and being directly involved in a process is crossed. For instance, the association of podzolisation with a particular vegetation community or even with a single tree, as in the case of the 'egg-cup' podzols of New Zealand (Bloomfield 1953), is dependent upon the leachability sequence being altered by organic matter under a particular set of conditions (see p. 74). At this level vegetation is not determining the balance achieved between various processes, as it would if acting as a factor, but is directly involved as an integral part of a process.

It is virtually impossible to deal with animal life in a factorial manner, even to the limited extent possible in the case of vegetation, for animals are so involved in complex feedback processes that any attempt to consider animal activity in soil formation must be made within the framework of these complex interactions, that is at the processes, rather than the factorial, level.

The position of the biosphere as a factor of soil formation is thus seen to be rather equivocal for it occupies a considerable range of influence, from an intimate involvement with processes at one end to something approaching factorial status at the other.

Time

Time as a factor in soil formation occupies a special position in that not only do the processes of soil formation operate in a temporal framework, but the other factors also change through time, both in themselves and in their mutual relationships. This duality, together with the need to consider various time scales in the context of soil formation, warrants an extended discussion.

Despite the fact that no process of soil formation can realistically be dissociated from its temporal dimension, the number of examples in which the time dependence of processes has been established is fairly small. This situation arises because it is difficult to establish an absolute time scale of sufficient length in relation to the speed at which the surface processes operate. Attempts have been made to solve this problem by field experiments over a number of years, and then extrapolating over the required time period. Thus Hilger's work on rock disintegration, as quoted by Jenny (1941), is typical. In this research uniform rock particles of limestone, schist and sandstone from 10–20 mm in diameter were exposed to the atmosphere and their disintegration observed over a period of 17 years. However, it is not obvious what can be obtained from the results other than a relative measure of rock disintegration for these particular rock types within a particular region. The same comments can be made on a more popular variant of this method in which tombstones and building rocks are used as indicators of the rate of disintegration. Many examples of this, covering a variety of rocks in a variety of climates, are given in Jenny (1941) and Ollier (1969).

With regard to the lateral processes of surface movement, reference was made in Chapter 6 to experiments to measure the speed of wash and

creep (Carson and Kirkby 1972). The extremely small annual incre-
ments recorded over a relatively few years by these methods is a very
unsure base for extrapolating to periods of thousands of years. Exactly
the same comments can be made regarding denudation rates derived
from only a few years' recording of the solid and dissolved load carried
by major rivers.

A time scale can be established by making use of a dated catastrophic
event, whether this is natural or man-made, as marking the time of
initiation of surface processes. Thus, Van Baren (1931) investigated the
rate of soil formation on the volcanic ashes deposited from the Kraka-
toa eruption of 1883. 1876, the year in which Lake Ragunda was drained,
was used by Tamm (1920) as the time when soil formation started on
the lake-bed sediments. Salisbury (1925), in studying a dune sequence at
Southport, England, used multiple lines of evidence, rather than a
single event, to establish a time scale, as did Dickson and Crocker
(1953) for the mudflow sequence of Mt Shasta, California. Hissink
(1938) studied the changes that have taken place in polder soils in
Holland over the past 300 years. This, of course, is a man-made datum
determined by the draining of a particular polder. More recently,
Paton *et al.* (1976) have reported on the rate of profile development in
coastal sands subsequent to heavy mineral extraction. In this case the
initiation of surface processes can be dated very accurately.

The ability to date materials, particularly those of the last one million
years, by radiometric means has provided a much more reliable and
extensive time scale for surface process studies. This can be seen in the
work of Fieldes (1955) in New Zealand, Hay (1959) in the West Indies
and Ruxton (1966) in New Guinea. It should be noted in all these cases
that dated materials are volcanic ash deposits.

When the processes investigated against these time scales are considered, it is apparent that for the most part they are those which are
relatively superficial and easily observed, such as the leaching of
calcium carbonate and associated variations in pH (Salisbury 1925,
Hissink 1938). In terms of the overall view of pedogenic processes given
in Chapters 1 to 8, the build-up of organic matter in the topsoil,
central to many investigations (Van Baren 1931, Dickson and Crocker
1953), must also be regarded as being relatively peripheral, even though
to plant ecologists and agronomists it is central to the problem of
nutrition. The one type of profile development that has been investigated to any extent against a time scale is that of the podzol. However,
it was shown in Chapter 7 that the development of this profile is a special
case dependent upon the accumulation of an inert quartz sand in association with a particular type of organic breakdown product and,

since it is capable of developing very quickly, it cannot be used as a general model of profile formation. It was only with the introduction of more sophisticated, physically determinative methods in the 1950s that the time dependence of epimorphism began to be unravelled. A notable example is provided by Fieldes (1955), which was discussed in some detail in Chapter 4 (p. 40). To date there has been no commensurate breakthrough in the investigation of the processes of lateral movement.

Taking an overall view, including even the most recent developments, investigations of time as a factor in pedogenesis have concentrated on processes that are essentially superficial or those which operate at a considerable speed. The time scale involved in more fundamental re-actions, such as the speed of epimorphism in granites or the rate of operation of surface processes, is not really known. Yet it is largely on the basis of work on organic matter accumulation and carbonate leaching that the concept of the steady state and soil maturity depends. In view of the previous comments, it seems to be a very unsure foundation for such a fundamental postulate.

Up to this point the passage of time has been dealt with in relation to pedological processes, what may be called 'pedological time'. However, when geological time is considered another question needs to be answered: What changes, if any, have occurred in pedological processes throughout geological time? It is evident from the nature of some of the oldest sedimentary rocks exposed in North America, Europe, South America and South Africa that water, wind and ice were operating at the lithosphere surface in early PreCambrian times in much the same way as they do now. In gross terms this means that most of the processes discussed previously in Chapters 1 to 6 have been active for at least the last 3000 million years. There is, however, a major exception, a consequence of the gradual development of life on earth, which must have resulted in considerable variations in the balance achieved between the various surface processes within that time period.

The initial establishment of life forms must have occurred under reducing conditions (Bernal 1967), which means the absence of oxygen. An important consequence is that at the surface of the lithosphere iron would have occurred in only the ferrous form; that is, as a highly mobile ion of group I (Fig. 3.1). At present, iron in the ferric state is one of the important components contributing to the stability of a great number of soil fabrics; its absence would have caused the degree of soil fabric stability to be much lower than it is today and there would also have been a general lack of red, brown and yellow colours in the surface zone of the lithosphere. The general weakness of fabric development would have allowed the process of sorting according to particle size to occur with much greater ease, making the accumulation of quartz sand on the land areas of the world relatively more common. The lack of ferric iron persisted until green-plant photosynthesisers developed some 2000–1800 million years ago (Cloud 1968). At this point ferric iron became a stable species in the near-surface environment, which is reflected in the first appearance of 'redbeds' (detrital continental or marginal marine sediments in which individual grains are coated with ferric oxides). From this time onward, many materials occurring on the land surfaces of the world would have acquired a somewhat greater degree of fabric stability and so the ease of sorting would have been reduced.

The development of terrestrial vegetation in the late Silurian, 420 million years ago, consequent upon the continued rise in the level of oxygen in the atmosphere (Berkner and Marshall 1965), made possible the initiation of many of the processes described in Chapter 7. In particular, this meant that clay–organic complexes could form and that there was a near-surface concentration of a wide range of elements and possibly podzol formation. The vegetational cover would also have impeded the freedom of surface movement of material and so made the process of sorting rather more difficult. Grasses first appeared 120 million years ago in the Cretaceous (Barnard 1964), but turf-making grasses did not appear until after the Miocene, about 10 million years ago. It is only since then that all the major processes of soil formation have been operating; consequently, for the greater part of the last 3000 million years the overall balance of pedological processes has been somewhat different from that which obtains today. This is of considerable significance when framing criteria for the recognition of palaeosols and when evaluating the degree of inheritance in modern soil formation.

It is apparent that, with the passage of geological time, the balance between the other factors must vary. This is most evident in the case of climate and vegetation, for instance the great changes which occurred during the repeated glaciations of the Quaternary. In terms of much longer periods of time, topography (the distribution of plainlands and fold mountains of the world) and the distribution of various major rock types, such as granite, basalt and sedimentary rocks, has altered considerably. However, our knowledge of past conditions is still very limited and precise statements about past changes in the factors of soil formation necessarily depend upon future research.

Conclusion

The preceding discussion supports the suggestion of Birkeland (1974) that the factors of soil formation operate at a higher categorical level than the processes. It must be emphasised, however, that there is no implication in the discussion that any factor acts in isolation, for they all operate together and set limits to the operation of processes as a whole. The discussion of each particular factor concentrated on certain situations in which that factor's control of processes is dominant and the effect of this control could be clearly distinguished. Thus, particular kinds of lithospheric material have a very considerable effect on the process of weathering, but such weathering cannot be dissociated from the simultaneous effect of all other factors.

The picture that emerges from this discussion is that the formation of soil material is the result of a complex of processes operating within boundaries determined by an interacting set of factors, the processes and boundaries being subject to variation during time. Many pedologists, in attempting to solve this complex problem, have tried to do so from the factorial viewpoint (Jenny 1941). It was realised that the factors were part of a complex system, but those situations in which certain factors played a controlling role, as discussed previously, were interpreted as being situations in which that particular factor acted as an independent variable with respect to the other factors, which remained more or less constant. The actual establishment of a factor as an independent variable (i.e. its separation from the factor complex) was largely achieved by the way they were defined by Jenny (Crocker 1952). The next step was to correlate the independently variable factor with certain characteristics of the soil material. It was thought that the overall process of soil formation would be understood through the accumulation of many examples of such correlations. Such an approach ignored completely the essential intermediate role of processes between factors on the one hand and soil material on the other. The reason for this situation is the division within pedology between those who deal with the place of factors in soil genesis and others who concentrate on the mechanisms of the actual processes involved. The boundary between these two approaches has been crossed very rarely (Crocker 1952). The correlations that have been achieved by the use of this factorial approach have been no more than the most simplistic or superficial. In essence this method, while partially recognising the complexities involved, attempted to solve the problem by retreating from the complex situation to a highly simplified one.

To solve the problem of soil formation it is necessary to acknowledge the full complexity of the problem and attempt to solve it in a holistic, rather than a particularate, manner. Central to such an attempt is an understanding of the relationship between factors, processes and soil material, as discussed in this chapter; the complexities in the processes of soil formation, as developed in Chapters 1 to 8; and, finally, the combinations that can occur between the factors, which will be the subject of the next chapter.

10

Pedological Provinces

The differentiation of pedological provinces

In the previous chapter it was seen that in any given situation the processes of soil formation are governed by the factorial complex as a whole. In certain areas, however, one or other of the factors departs markedly from what can be regarded as the 'norm' and, in so doing, it becomes crucial in determining which processes of soil formation are dominant and the balance that can be achieved between them. This forms the basis for the recognition of **pedological provinces**. It needs to be reiterated, however, that even in these rather special situations the particular factor which is regarded as being aberrant is not being dealt with as an independent variable, but as one which has caused the node of the factorial complex to become somewhat skewed.

Province 1 The most obvious of these deviant situations is that associated with the topographic factor in the fold mountain belts of the world. The generally steep topography within those areas means that the dominant processes of soil formation will be those of incomplete epimorphism and lateral mass movement (p. 100). However, the effect of associated factorial changes, particularly those of climate and lithospheric material, must be taken into account if the factorial control of the processes of soil formation within the fold mountain province is to be fully appreciated.

In areas of high altitude generally associated with fold mountains precipitation is mainly in the form of snow, and hence glaciation and all the associated forms of ice action are common (see p. 64). Fold mountains are also commonly associated with extremely explosive vulcanism, which is responsible for the continuous addition to the surface of considerable amounts of volcanic ashes and lavas. The rapid transfer of this material from magma chamber to the surface means that a number of igneous rock minerals are in a highly disordered state or even in the form of glass. In this situation epimorphism takes place at considerable speed and this results in turn in the production of clay minerals with disordered structures of the allophane type, giving soil materials some unique properties. The speed of epimorphism is attested to by the fact that most recent work on the time factor in pedogenesis has been done on volcanic ash from New Zealand, New Guinea and the West Indies (p. 104). Lateral processes of movement, particularly mass movement, are also affected by the addition of this unconsolidated material to the lithosphere surface, a tendency which is further augmented when account is taken of the earthquakes and actual earth movements which are relatively common within the fold mountains.

Province 2 Turning attention to the plainlands of the world, one can recognise three other areas in which certain of the factors of soil formation are markedly deviant. The most readily demarcated of these areas is that part of the northern hemisphere, both in Eurasia and North America, which has been subject to repeated glaciations in the Quaternary. The lithospheric material inherited from these glaciations is unconsolidated and extremely varied both in grain size and mineralogy, although the degree of sorting varies from the poorly sorted morainic material that has been derived directly from glacial processes to the much better sorted materials that have been subject to subsequent action by water or wind, such as outwash deposits and loess. The deposition of these materials has given rise to subdued but irregular topography with disrupted drainage systems, particularly in lowland areas.

When these factors of lithospheric material and topography are considered together with the short time period involved since the end of the Quaternary glaciations, it is obvious that the degree of inheritance possessed by soil materials of this province must be very much greater than elsewhere in the world's plainlands and hence a very distinctive assemblage of soil materials must result. As a direct consequence, most epimorphism is secondary, and lateral processes of surface movement are not of great importance. Because of this, the pedological processes that have been investigated have been relatively superficial, such as gleying, the movement of carbonate, and the accumulation of organic matter. The only case of fundamental profile development investigated is that of the podzol, which has been able to assume a marked character in the time available, because of the speed at which the biospherically controlled processes can operate in outwash bodies of quartz-sand (see p. 74).

Province 3   The factor of lithospheric material can again be regarded as aberrant when the rocks are of a basaltic nature, accepting granitic type material as the norm. The processes of epimorphism are affected to a considerable extent, for the lack of highly resistant minerals ensures that the final product is of uniformly fine-grained newly formed minerals. Such minerals are combined to form some very stable fabrics resistant to mechanical breakdown by rainfall impact and movement due to surface wash, which means that lateral processes are dominated by creep and, in certain circumstances, by slippage.

Province 4   A fourth pedological province can be recognised within those areas of the world's plainlands where vegetation is sparse or non-existent and the annual low rainfall arrives in the form of short intense downpours. The deviant nature of the processes of soil formation must be ascribed to the joint effect of these two factors, which means that the separate discussion of these two topics (see pp. 101 and 102) must be combined. The processes of epimorphism in such areas can be assessed only in terms of the end products, i.e. the newly formed minerals, which are almost confined to lower topographic situations such as playas. Their mineralogy seems to indicate that weathering, leaching and new mineral formation are going on in much the same way as in a more humid areas but no doubt in a rather more episodic manner. In addition, there is a greater tendency for highly mobile ions such as chloride, sulphate, carbonate, sodium and magnesium to accumulate in lower topographic situations and hence influence the character of the soil material. The combination of lack of vegetation and intensity of rainfall causes lateral movement to occur with much greater speed than in more humid, vegetated areas. Wind is also important as an agent of transport and concomitant sorting. Such differences from the processes of soil formation in other areas within the world's plainlands are sufficient to give rise to a distinctive suite of soil materials.

Province 5   Up to this point four pedological provinces have been differentiated according to the 'abnormal' nature of one or more factors of soil formation. This leaves a core area, which forms the fifth– and most extensive – province covering great areas of Australia, Africa, peninsular India and South America, where by implication the factors are 'normal'. It is, however, much more difficult to define normality than to define a factor which is markedly deviant. One possibility is to define the fifth province in a negative sense as being controlled by a factorial complex, none of whose elements deviate in the manner specified previously for the other provinces. Another method is to define the factorial complex within this province in a positive way; this means defining each factor instead of only one or two, resulting in a series of broad statements such as: the climate is reasonably humid and the vegetation cover adequate; lithospheric material is of a granitic type or its sedimentary equivalent and topography is gentle enough so that residual, transportational and depositional sites can be differentiated.

Vague as these statements of factorial control are, they are sufficient to indicate the balance achieved between the various processes of soil formation within this province. For, given the long-term relative stability of such land areas in combination with sufficient humidity, the processes of epimorphism have the greatest opportunity to approach their end points. Lateral processes of movement are those in which water is dominant and, if sorting according to particle-size is possible, it occurs but relatively slowly because of the intervention of vegetation.

The soil materials of the pedological provinces

The discussion of soil materials of these five pedological provinces will be limited to those which occur in provinces 3, 4 and 5, for, while the importance of the marked differences of the soil materials of the glacially influenced plainlands of province 2 and of the fold mountains of province 1 are realised at the level set out on page 109, the author does not possess sufficiently detailed knowledge to express these generalities in specific soil material terms.

Pedological provinces 3, 4 and 5 together encompass a closely interrelated group of soil materials developed within the world's plainlands. Descriptions of the soil materials of province 5 (the core area) will be given first for they can be considered as types to be compared with the materials of province 3 and finally of province 4. These descriptions are not intended to be a complete coverage of all soil materials but rather a selection of important nodal areas from which a general understanding of the relationship between factor, process and material can be obtained within these three pedological provinces. The basis for soil description is mineralogy and fabric, for both of which the necessary range of terms has been covered in Chapters 1 to 8. There are certain other terms which will be used dealing with field texture grading and consistence; an explanation of their use is given in the Appendix. Despite the fact that many of the following soil descriptions will be given in the form of profiles, there is no implication that these profiles are necessarily considered as genetic units, i.e. formed by pedological processes operating vertically downwards from the surface for, as was demonstrated in Chapter 6, there are many situations, particularly on transportational and depositional sites, where this is not the case. The profile is merely a convenient method of summarising a great deal of pedological information and a ready means of correlation in certain circumstances.

The soil materials of the fifth province   As a start consideration will be given to soil materials that result from a granite being subjected to maximum epimorphism and lateral surface processes dominated by water. This will then allow the more complex situation of soil material derived from sedimentary rocks under similar conditions to be evaluated.

Soil materials derived from granite   The mineralogy that results from the impact of epimorphic processes over a considerable time period is a mixture of residual sand-sized quartz grains and clay-sized kaolinitic materials together with oxides and hydroxides, mainly of ferric iron. The fabrics developed are determined by the degree to which these minerals are affected by lateral processes and can be rationalised in terms of materials associated with residual, transportational and depositional sites (Fig. 9.1). In general, it is possible to describe these materials at residual and transportational sites in terms of profiles, but this is not usually the case for depositional sites.

At residual sites the characteristic profile, in terms of field texture grade, shows a gradual change with depth, from a sandy loam only a few centimetres thick, to a sandy clay loam and then to a sandy light to medium clay, which generally extends to about a metre in depth. Below this level the degree of porosity, and the subplasticity, can decrease markedly with closer packing of the sand grains.

The near-surface tendency to single-grain fabric and lighter texture grade can be ascribed to fabric breakdown by rainfall impact and the removal of finer particles by pervection (p. 59). It should be particularly noted that this process has not been responsible for any sharp junctions within the profile, but only for gradual change with increasing depth. Profiles such as these can be referred to as **residual earths**.

Transportational sites are typified by profiles that show an abrupt contrast between a surface layer, dominated by processes of lateral movement, and underlying material derived from bedrock by epimorphic processes. The surface layer, which is between 20–30 cm thick, has a field texture grade of loamy sand to sandy loam. The matrix fabric is non-uniform in that it consists of areas, a few millimetres apart, that are alternately closely packed and highly porous; the pores are also highly irregular, both in size and form. Planar voids are absent or at most very sporadically developed. The consistence is very brittle and fragile, breaking down to single-grain sand with ease. The fabric of this material is the same as that described for depositional sand bodies (p. 56) and can be referred to as **secondary earthy fabric**, compared to the primary earthy fabric of the residual earths. Because of the ease with which this fabric can be broken down, the surface, often to a depth of several centimetres, has a single-grain fabric owing to rainfall impact and surface wash.

The lower boundary of this sandy loam surface layer is well defined and may also be marked by a stone line (Fig. 6.3). The underlying material is equivalent to that which occurs at depth in the residual earth profile and consists of a very sandy clay with few, if any, planar voids. In most cases the matrix fabric is somewhat porous and a bolus (see Appendix) can be formed with ease from the resulting friable material. However, where alkaline granites are involved, the porosity decreases so that the material becomes much more resistant and bolus formation is difficult. The general nature of the first material can be described as *mellow* and the second as *harsh* and the resulting profiles as mellow texture contrast soils and harsh texture contrast soils, respectively. These two types of profile will be dealt with in greater detail in the next section when the derivation of soil materials from sedimentary rocks is considered. Also, it will be more convenient to postpone considerations of the soil materials associated with depositional sites until the end of this next section.

Soil materials derived from sandstone/shale sequences Where, as is common, the dip of the strata is low, the topography is often controlled by a massive quartz-rich sandstone and finer-grained members of the sequence are restricted in outcrop to the hillslopes (Fig. 10.1). The first consequence of this arrangement is that there are widespread plateau surfaces which are well demarcated residual sites, within which the dominant epimorphic processes acting on the quartz-rich bedrock form a residual earth profile. This profile is very like that derived from granite but differs most particularly in that the lower boundary is frequently abrupt, being determined by the boundary of a particular sandstone bed.

In the case of transportational sites on hillslopes, epimorphic processes operate on the finer-grained bedrock to produce pockets of clay-rich materials with highly developed planar void patterns, for the most part inherited from the bedding and jointing pattern of the bedrock. The overall pattern of soil profile development in such transportational sites is thus rather different from that described for granite: the pattern consists of a mixture of texture contrast profiles where the mobile surface layer with secondary earthy fabric overlies the clay-rich highly pedal materials, interspersed with shallow uniform sandy profiles where the mobile surface layer directly overlies more resistant sandstone bedrock (Fig. 10.1).

There are mellow and harsh types of these highly pedal subsoil clays just as with materials derived from granite. The mellow highly pedal clays have a good development of subdominant minor voids and a somewhat porous matrix fabric. In the moist state such materials are friable, but when dry they are firm. In contrast, the harsh, highly pedal clays have little if any development of subdominant minor voids and a very dense matrix fabric. In the moist state they are very sticky but very hard when dry.

Not only are these differences important in themselves, but the effect that they have on the secondary earthy surface layer is very considerable. The character of the harsh clay is determined by the ease with which its constituent clay particles assume the deflocculated state and this in turn affects the way in which water moves through the soil. Given the secondary earthy fabric of the surface layer, water will move through such material with comparative ease along many routes. When the harsh clay layer is reached, further downward movement will be almost wholly restricted to the fairly coarse pattern of planar voids: the concentration of water flow along these planes will lead to a quick deflocculation of the ped surfaces adjacent to these voids. Deposition of these deflocculated particles in a short distance effectively prevents any further movement of water along these voids. Thus, where harsh clay subsoils occur, even a fairly light shower of rain may be sufficient to cause an almost watertight seal to form near its upper surface; this will

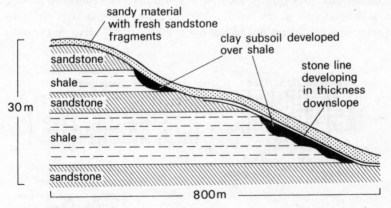

Figure 10.1 Texture contrast soils, transportational site, shale/sandstone sequence.

persist until subsequent drying and contraction of the deflocculated clay re-establishes the continuity of the planar voids. Thus, over a time period such as a year the base of the sandy layer will be subject to numerous short periods of intense waterlogging, resulting in the development of a thin but intense bleached zone which is separated from the underlying clay by a very sharp boundary. The shape of this boundary is determined by the nature of the planar void pattern of the clay layer. For example, in some clays there is a subdominant major void pattern giving rise to a rather weak development of prismatic peds (Fig. 10.2(a)).

In contrast, where mellow clays are involved, the finer dominant planar void pattern, the common occurrence of subdominant minor voids and the more porous matrix fabric, which are all a reflection of the more strongly flocculated nature of the clay, allow water to penetrate more easily into and through them without interruption. Impeded drainage can only occur after prolonged rainfall, say in response to wet season conditions, and probably in lower topographic situations. In the surface layer this produces a wide zone of bleaching which is neither so intense nor so sharply demarcated as when it overlies a harsh clay subsoil. Furthermore, this is accompanied by mottling in the clay subsoil. Because of the lack of deflocculation at the boundary between the surface layer and the subsoil, the boundary is not so sharply demarcated as is the case with the harsh clay subsoils and, even if subdominant major planar voids define a prismatic structure, there is no development of columns along the boundary.

This discussion of the effects on the surface layer can be applied with only a few amendments to the case of mellow and harsh texture contrast soils derived from granite, the only major difference being their lack of a well-developed planar void pattern. Furthermore, it is possible to apply these considerations of soil materials on transportational sites to

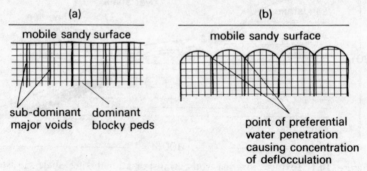

Figure 10.2   Development of columns in subsoil clay.

soils derived from a much wider spectrum of sedimentary bedrock. It has been found that, in many cases, what are regarded by geologists as pure sandstones or pure shales generally contain sufficient clay-sized particles in the first case or resistant sand-sized particles in the second to make the generation of texture contrast soils possible in both instances. The sandstone-derived profiles can approach those previously described for granite in displaying little pedality in the subsoil, and the shale-derived profiles frequently have only a thin mobile surface layer which, in many cases, has such a fine sand grain-size that it assumes a dense closely packed matrix fabric.

Up to this point only gently dipping strata have been considered. Where the dip is much greater, there are some important amendments to the scheme discussed above. The chief of these is that residual sites are less extensive and transportational sites are more important. A typical situation is shown in Figure 10.3 where soil materials are shown developed on a steeply dipping succession of quartzites and shales. Here, the mobile surface layer is continuous right across the crest, except where it is interrupted by the outcropping quartzite, for there is practically no point where lateral processes of surface movement are not operating at a considerable speed. The strongly resistant quartzite contributes many large joint demarcated blocks to this surface layer, so that it consists of a mixture of very coarse cobbles and gravels in a matrix of sandy loam with a secondary earthy fabric. These very coarse fragments are concentrated at the base of this surface mobile layer as a very prominent stone line. The development of a clay subsoil is restricted to the interbedded shales and can be harsh or mellow in its general character.

The whole range of soil materials derived from granites and sand-

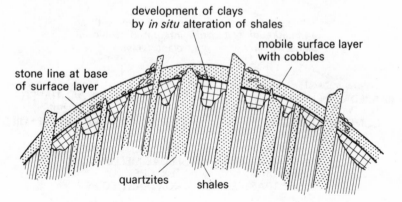

development of clays
by *in situ* alteration of shales

mobile surface layer
with cobbles

stone line at base
of surface layer

quartzites   shales

Figure 10.3   Soil development on steeply dipping sediments.

stones and shales on transportational sites generates a great number of profiles that differ from one another in certain of their characteristics. Because the previous discussion has shown that many of these differences are dependent on whether the subsoil is mellow or harsh and whether or not a planar void pattern is well developed, it is possible to rationalise the great variety into four nodal types (Fig. 10.4):

Depositional sites are characterised by the sorted products of lateral movement and the resulting materials are either dominantly sands or clays. The sandy materials have a secondary earthy fabric of the same kind as that described for the surface layer of the texture contrast soils and may be referred to as **depositional earths**. The more clay-rich materials have a much denser fabric than when clays are developed *in situ*. At the same time, although the planar void pattern is well developed, its dominant mode is much coarser and the subdominant minor pattern is developed to a lesser degree (see p. 51). This change in fabric means that in general these depositional clays are somewhat harsher than clays formed *in situ*, but even so it is possible to distinguish, within the range covered by such clays, those which can be regarded as mellow from those which are harsh.

These contrasted materials frequently form interbedded deposits and in these circumstances it serves no useful purpose to attempt to describe them in terms of profiles. It is best to consider them as lithological units with particular fabrics forming part of a sedimentary sequence. In certain circumstances, such as those in Figure 6.2, large segregations of relatively uniform depositional earths and depositional clays can occur.

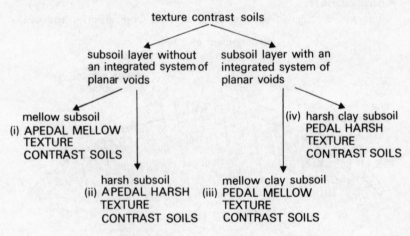

Figure 10.4 Texture contrast soils.

In such situations these materials can develop certain characteristics which it is possible to consider in terms of profiles.

If depositional earths are deep enough and a particular vegetation type becomes established on them, a **podzol** can develop (p. 74). This is a common occurrence in footslope positions where the coarser sand fraction is preferentially deposited. These profiles consist of a surface layer a few centimetres thick of single-grained, organic, stained, loamy sand which grades into a close-packed, brittle, bleached sand, up to 20 or 30 cm thick. This is underlain, across a very sharp, highly irregular and convoluted boundary, by a double pan, one in which organic matter is concentrated in a zone 2 or 3 cm thick near the boundary and another, lower down, in which iron is deposited, which is more sporadically developed over a thickness of 20 to 30 cm. The pan is indurated, requiring a hammer blow for its disintegration when dry, and is highly brittle. The field texture grade of the pan is a loamy sand and this same texture grade continues into the unindurated material below. In terms of fabric there is a contrast on either side of the pan: single-grain, close-packed above and secondary earthy both within the pan and below it. It seems that the process of podzolisation, as well as stripping iron from around the quartz grains, has also destroyed the secondary earthy fabric.

In those cases where depositional clays have a depth greater than about 50 cm, a new type of planar void pattern develops in which the planes of weakness are at an angle of about 45° to the surface, probably in response to loading pressure. This oblique pattern of planar voids can be very well developed and define parallelpiped-shaped peds that can vary in size along their major horizontal axis from a centimetre to a metre or more.

While for many purposes it may be best to deal with lithological units within depositional areas, for purposes of comparison with other nomenclatures it is necessary to speak in terms of profiles. As has been demonstrated, it is possible to do this when the sands and clays assume a reasonable thickness. In these circumstances it is possible to distinguish:

    (i) Depositional earths,
    (ii) Podzols,
    (iii) Mellow depositional clays,
    (iv) Harsh depositional clays.

The soil materials of the third province   Soil materials derived from the epimorphism of basalts are uniformly fine-grained, but are of two distinct types:

(a) Red to brown materials in which kaolin and oxides, mainly of iron, are dominant,

(b) Dark-coloured materials in which 2 : 1 clays are dominant.

The formation of these two types of material can be related to different balances between the various epimorphic processes (Fig. 4.4). It has been common to relate such a difference to present-day macroclimatic factors (Hallsworth 1951; Singer 1966) or to the time factor, i.e. the 2 : 1 clays are formed first and then, with the passage of time and the further action of epimorphic processes, break down to kaolin and oxides occur (Ferguson 1954; Nicolls and Tucker 1956). In certain areas, however, these two materials occur in such close association, that other factors besides climate must be responsible for the difference; furthermore, an evaluation of sites at which these two materials occur in terms of the impact of epimorphic processes and those of lateral surface movement suggests there is no basis for invoking a time difference between them. Beckmann *et al.* (1974) discuss such an occurrence from the Darling Downs of southern Queensland. It was shown that the brightly-coloured, reddish–brown, kaolin/oxide-rich materials are restricted to the surface of mesa-like hills formed by the outcrop of a resistant basalt, while the dark-coloured 2:1 dominated clays occur on the pediment-like slopes developed on a whole range of much less resistant basalts which extend outwards from the isolated hill features towards the vague drainage lines. The hill tops are clearly residual sites at which epimorphic processes can operate intensively with little or no intervention from processes of lateral surface movement, while the long gentle hillslopes are transportational sites on which the dark-coloured clays form a mobile surface layer. The evidence for this has already been discussed on page 62 and also in Paton (1974).

The reddish–brown, kaolin/oxide-rich soils of the residual sites vary in field texture grade from clay loam or light clay at the surface to medium clay in the subsoil. The matrix fabric is rather porous and the planar void pattern is a well-developed fine orthogonal one giving rise to 2–3 cm blocky peds with numerous subdominant minor planar voids of the same type, which ensures a ready breakdown to much finer entities. This type of fabric determines the friable to firm consistence and the marked subplasticity of the material. The depth to basalt varies from 30 to 80 cm and slabs of unweathered basalt are common throughout the profile. Profiles such as these can be referred to as **subplastic residual clays**.

The very dark brown, medium to heavy, 2 : 1 dominant clay is 30–50 cm thick. The matrix fabric is very dense and there is a well-developed orthogonal planar void pattern giving rise to blocky peds which vary in size from something less than a centimetre at the surface to 5 cm or more at depth. There is, however, only a very slight development of subdominant minor planar void patterns within these peds. The consistence is very firm to hard when dry and highly plastic to somewhat sticky in the bolus form. In many places this layer of material directly overlies basalt across a sharp boundary. Such profiles can be referred to as **dark-coloured blocky clays**. In other places, however, particularly in downslope positions, this dark clay layer overlies a material which in terms of fabric and colour is in strong contrast. It is a much brighter, yellowish– or reddish–brown, medium clay with a dense matrix fabric and a well-developed, fine, oblique, planar void pattern. This void pattern gives rise to irregular wedge-shaped peds of from 5 to 10 cm in size. In addition, there are many subdominant minor oblique planar voids which enable the dominant peds to be broken down fairly readily into finer entities. Carbonates are generally common both as nodules and in much finer-grained form. The material has a firm consistence when dry, but is friable when moist, and a bolus can be formed with considerable ease. This material, together with the overlying dark clay layer, may be considered to be a **fabric contrast soil**.

These transportational sites on the long gentle hillslopes with their mixture of dark-coloured blocky clays and fabric contrast soils merge very gradually into depositional sites on valley floors. The soil material of such sites is naturally of a clay texture only. Again, as in the case of the soil materials of province 5 in similar sites, it is difficult to talk in terms of profiles because of the bedded nature of the deposits. However, where an individual layer is of a sufficient thickness, profiles can be differentiated that have the same characteristics as those described on page 119 for depositional clays. Because of the material from which they are derived, the greater number of such profiles would be considered as harsh depositional clays with a surface which develops deep major cracks during the dry season.

It is possible to compare this general sequence of soils developed at residual, transportational and depositional sites with those developed within the fifth province, as shown in Figure 10.5.

The restricted occurrence of the subplastic clays, in the particular example used, points to the fact that this is very nearly a limiting situation for their production. It is possible to extrapolate from this rather special case into areas where the balance between the epimorphic processes is such that the production of 1 : 1 clays and free oxides is much more generalised. In these situations the material maintains its subplastic, highly pedal nature, although the peds are normally polyhedral in shape rather than blocky and the profiles may be 2–3 m deep. The most important difference is in the range of sites over which this material is developed, for it is found not only on residual plateau surfaces but also on steep hillsides which would normally be regarded as transportational sites. It would appear that the strength cf development of the fabric in these subplastic, highly pedal materials is such that the processes of lateral movement, apart from mass movement, are nullified to a large extent.

The soil materials of the fourth province   The distinctive character of this province resides in the rather different way in which soil material is distributed across the landscape. The lack of vegetation means that when rainfall does occur it considerably reduces the possibility of soil material accumulating in either residual or transportational sites. Therefore, bare rock surfaces are much more common in the higher parts of the landscape, although where soil materials do occur they are of the same general types as those described for provinces 3 and 5. It should be emphasised that there are no soil materials associated with residual and transportational sites that are unique to province 4.

There is, of course, a relative concentration of soil material in depositional sites. In purely sedimentological terms the materials are very much the same as those of province 5, consisting of sorted sands and clays which are generally interbedded but which, on occasions, may be segregated into uniform bodies of considerable dimensions. Within such segregated bodies of sand it is possible to recognise two main types: one is equivalent to the depositional earths described previously and, like them, has a secondary earthy fabric; the other type is confined to such sands as have been subject to the winnowing action of the wind which effectively removes most of the fines responsible for the secondary earthy fabric and results in single-grain sands which do not cohere to any extent. Such materials can be referred to as **depositional sands**.

	3rd Province	*5th Province*
residual sites	sub-plastic residual clay	residual earth
transportational sites	dark coloured blocky clay	mobile surface layer with secondary earthy fabric
	fabric contrast soils	mellow and harsh texture contrast soils
depositional sites	harsh depositional clays	mellow and harsh depositional clays

Figure 10.5 Soil/site comparison for province 3 and province 5.

Any segregated bodies of clay are essentially the same as the depositional clays described previously. Because of the accumulation of the more mobile cations, such as sodium and magnesium, which results from the lack of through drainage, most of these clays have extremely harsh characteristics.

The general lack of through drainage is also responsible for the high concentration of mobile anions such as carbonates, sulphates and chlorides, at least within depositional sites. This is in contrast to the occurrence of these anions in provinces 3 and 5 where the few concentrations that do occur are associated with particular bedrocks and are quickly diluted in the course of movement away from their source. The higher mobility of chloride and sulphate leads to an almost complete concentration in depositional sites, particularly in clays where they often form a surface crust. On the other hand, carbonates, being relatively less mobile, occur as either extremely fine grains, nodules or more massive deposits, in association with a wide range of soil materials. Once the status of this carbonate is recognised as being essentially secondary to that of the soil material, it follows that it is unnecessary to differentiate as unique every association between carbonates and a particular soil material as has been done by many pedologists (Stace *et al.* 1968). The differentiation of palaeosols is another problem that is particularly associated with province 4. Until more is known about the speed at which particular pedological processes operate within this province over periods of at least a few centuries, it would seem rather unwise to ascribe the occurrence of a particular soil material to past climatic differences without a great deal of corroborative evidence from non-pedological sources.

The possibility of correlation with previously used soil nomenclatures

The difficulty of trying to correlate between the basic units described earlier in this chapter and those used in other systems is that in each scheme units are related to a different model of soil formation. These difficulties are augmented by the fact that many terms common to various classifications do not necessarily refer to the same kind of material (Paton and Williams 1972). The model that has been proposed in this book is markedly different from those which have been proposed previously, so that it is inevitable that considerable difficulties will occur when such a correlation is attempted. The range over which correlations can be attempted is immediately restricted by the concept that within each of the pedological provinces defined on page 109, there is a characteristic assemblage of soil materials. This means that the possibilities of correlation within a particular province are much greater than between provinces. The lack of recognition given to these fundamental restrictions has confused all attempts at soil classification. Because pedology developed almost entirely within the glacially influenced plainlands of province 2, the most influential model of soil formation was developed within these narrow confines and then applied to the rest of the world. This is illustrated by the soils and vegetation maps of Africa by Shantz and Marbut (1923) and of Australia by Prescott (1931). In these publications the soils were made to fit a model in which it was accepted as axiomatic that there was a relationship between climate and vegetation on the one hand and characteristic soil profiles on the other, and that this relationship had a worldwide applicability, i.e. world zones could be demarcated.

In attempting to correlate soil materials of provinces 3, 4 and 5, as described earlier, with those of other systems, it is essential to confine attention to those systems that deal with the soil material of these three provinces and, in particular, to be wary of the effects that zonal models derived from province 2 may have on such systems. Such a task is particularly difficult in the case of Africa where several different versions of the zonal model of province 2 were imported by the British, French, Belgians and Portuguese. This had led to an enormous tangle of over-lapping categories. Work has been concentrated on trying to rationalise this situation from a nomenclatural point of view (D'Hoore 1964), while detailed soil description, which is required if reliable correlation is to be achieved, has been largely ignored. In the case of Australia the problem of correlation is not so difficult, for the idea of explicit zonal controls has been abandoned for the most part; instead, there has been a great emphasis on the detailed description of soil material. However, care still has to be exercised, for zonal influences of the northern heimsphere are still implicit in the retention of a number of terms derived from the zonal model for the definition of soil categories (Stace *et al.* 1968). Even where these terms have been abandoned (Northcote 1971), the profile is still retained as the only possible basic unit and with it one of the basic zonal concepts: the notion of genesis related solely to vertically operating processes.

A result of this general acceptance of the profile as the only possible basic unit is that, if any correlation is to be established between the system of classification proposed in this book and previous systems, it must be couched in terms of profiles. This is why, earlier in this chapter, various 'type profiles' were demarcated even though in terms of the model developed in this book the more fundamental units are the various types of soil material. The following discussion will not be a full discussion of various classificatory schemes but only a means of relating, in a very generalised manner, the 'type profiles' defined earlier to previously recognised profiles. Comment will be restricted to the schemes of Stace *et al.* (1968) for Australian soils and of D'Hoore (1964) for African soils, as these are the most recent and most generalised statements on the soils of these two continents.

Australian soils The suggested correlation between the type profiles described earlier in this chapter and the Great Soil Groups of Stace *et al.* (1968) is given in Figure 10.6. It must be emphasised that there is in general no one-to-one correlation. For example, while the greater part of Stace's Red and Yellow Earths can be equated with the residual earths of the present scheme, there is a considerable overlap of these two groups into the depositional earths. Again, the way in which the Prairie Soils and Black Earths are related to the dark-coloured blocky clays and the fabric contrast soils is difficult to determine because fabric

Site	Type profile	Great soil group
Residual	residual earth	red earths yellow earths
	residual sub-plastic clays	kraznozems xanthozems euchrozems
Transportational	dark-coloured blocky clays	prairie soils shallow black earths
	fabric contrast soils	black earths chernozems
	mellow texture contrast soils	grey–brown podzolic soils red podzolic soils yellow podzolic soils brown podzolic soils lateritic podzolic soils gleyed podzolic soils
	harsh texture contrast soils	solonetz solodized solonetz and solodic soils soloths solonized brown soils red–brown earths non-calcic brown soils desert loams
Depositional	mellow depositional clays harsh depositional clays depositional earths depositional sands podzols	wiesenboden grey, brown and red clays earthy sands siliceous sands podzols

Figure 10.6 Suggested correlation of 'type profiles' and Australian Great Soil Groups.

contrast soils have hitherto not been recognised as such. However, the greatest difference lies in the texture contrast soils in which those that have a mellow clay subsoil are equated with the podzolic soils, and those having a harsh clay subsoil with the solodic soils. A considerable rationalisation has been achieved because the new model of formation suggested for these soils in Chapter 6 emphasises the primacy of texture contrast. The vertical processes of clay shift or differential clay destruction in the A horizon, invoked in the formation of Stace *et al.*'s podzolic and solodic soils, depend on a whole range of criteria both in the A and B horizons and are here regarded as secondary features. It is only in the case of podzols that there is an almost complete coincidence, for an extremely characteristic profile developed by vertically operating processes within a body of quartz sand is recognised by both systems of classification.

Even in the partial list of names used by Stace *et al.* (1968) and given in Figure 10.6 the influence of zonal concepts derived from province 2 may be seen. Some of these names, such as prairie soils, chernozem and wiesenboden, despite their soil group status, imply nothing more than vague similarity in terms of colour (prairie soils and chernozems) and topographic situation (wiesenboden). It must be emphasised these terms, as used by Stace *et al.* (1968) have a purely Australian connotation. The same, however, cannot be said for the podzolic and solodic groups for, in these cases, not only have the names been derived from the zonal model based on province 2, but the processes implied by these names have also been adopted.

The term **podzolic** as widely used implies that the processes of podzolisation are operating but have not proceeded far enough to produce a podzol. While such a relationship is established between the podzols and podzolics of province 2, this is clearly not true for province 5 where a completely different set of genetic processes operates to produce the mellow texture contrast soils. Application of the adjective 'podzolic' to these soils within province 5 has led to confusion and misunderstanding.

The terms **solonetz, solodised solonetz,** and **soloth** (Fig. 10.7), as widely used, imply the progressive vertical leaching of all free salts **(solonisation)** to produce a solonetz, followed by exchangeable sodium and magnesium **(solodisation)** to produce a solodised solonetz as an intermediate stage before the final soloth stage is reached. This viewpoint also implies an initial state where a uniform material is maintained in a high stage of flocculation by free salts. Such material has been called a **solonchak,** and a developmental sequence derived from it is said to occur around Lake Balaton in Hungary (De Sigmond 1938); however, it certainly does not apply within province 5 where the sharp texture contrast and the great variety of subsoil character can be explained in terms of lateral surface movement and bedrock character. A European influence can be seen in the use of the term **solonised brown soil** by Stace *et al.* (1968), i.e. a texture contrast soil produced by the process of solonisation. Before these ideas had penetrated from Europe, such soils had been called **mallee soils** because of the vegetation with which they are associated (Prescott 1931). The influence of this genetic model (based on sequential leaching) upon soil nomenclature within province 5 has been, and continues to be, very considerable; it seems to be an unnecessary complication.

Two other inherited names, **terra rossa** and **rendzina,** are not included in Figure 10.6 but need some comment, for, apart from indicating a reddish or a dark-coloured soil developed on limestone, they imply nothing more about the nature of the soil material. The soil materials included within these two categories are very widespread and none of them is unique to limestones. It would be better to describe the soil material in the terms outlined on page 112 *et seq.*, rather than in terms of bedrock which is not the primary determinant of the derived soil material.

African soils   In trying to correlate with the units of the map of African soils as defined by D'Hoore (1964) the great difficulty is that many of these units are defined in a non-pedological manner, for instance in terms of climate and bedrock. By eliminating such units from consideration it is possible to restrict discussion to five groups, as shown in Figure 10.8. However, even within these groups, there remain great difficulties in achieving correlations with the type profiles described

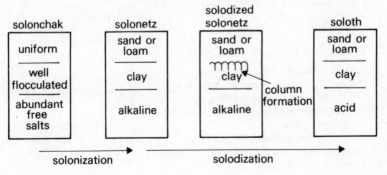

Figure 10.7 Development of solodic soils.

previously, because the major basis for D'Hoore's differentiation is the mineralogy of the soil material, which has very little relation with the gross morphology employed in this book in defining the type profiles.

With these provisos in mind, it is possible to say that the first division of the vertisols, those derived from rocks rich in ferro-magnesian minerals, can be equated with the dark-coloured blocky clays and the fabric contrast soils, while the third division, the vertisols associated with topographic depressions, contains equivalents to the harsh depositional clays. Within the halomorphic group, the solonetz and solodised solonetz are equivalent to the harsh texture contrast soils.

There are other factors that have to be taken into account when dealing with the other three groups: ferruginous tropical soils, ferrisols and ferrallitic soils. The first of these groups is a direct equivalent of the *sols ferrigineaux tropicaux* in the French system of Duchaufour (1960), which relates mineralogy to climatic zones, while the ferrisols and ferrallitic soils are equivalent to similarly named materials in the Belgian system as developed in the Congo (Sys 1960), in which considerations of profile play a much greater role.

Vertisols
1. Derived from rocks rich in ferromagnesian minerals
2. Derived from calcareous rocks
3. Of topographic depressions

Halomorphic soils
1. Solonetz and solodized solonetz
2. Saline soils, alkali soils and saline alkali soils
3. Soils of sebkas and chotts
4. Soils of lunettes

Ferruginous tropical soils
1. On sandy parent materials
2. On rocks rich in ferromagnesian minerals

Ferrisols
1. Humic
2. On rocks rich in ferromagnesian minerals

Ferrallitic soils
1. Dominant colour: yellowish brown
    (a) On loose sandy sediments
    (b) On more or less clayey sediments
2. Dominant colour: red
    (a) On loose sandy sediments
    (b) On rocks rich in ferromagnesian minerals
3. Humic ferrallitic soils

Figure 10.8 Part of legend for Soils Map of Africa.

Within both the ferruginous tropical soils and the ferrallitic soils, the major differentiation is based on whether the soil is dominantly sandy or clayey, and it is possible to say that certain of these sands and clays can be correlated with the residual earths and the residual subplastic clays, respectively. However, as D'Hoore's inclusion of particular soil materials within either of these groups is decided mainly on the basis of mineralogy, there are a great number of other morphological units included within them. Thus, in both the French and Belgian systems, but at a lower categorical level, a leached ferrallitic soil is recognised which can be equated with the mellow texture contrast soils. The ferrisols, however, can be equated almost wholly with the residual subplastic clays. It is of interest that D'Hoore (1968) found great difficulty in correlating the ferruginous tropical soils, ferrisols and ferrallitic soils with units in other systems such as the USDA's Seventh Approximation (Soil Survey Staff 1960), while his other units could be more readily correlated. D'Hoore ascribed this difficulty to the rather featureless nature of the soils and the many aberrant results from standard laboratory analyses. From what has been stated previously it would seem that these difficulties arise because the materials of these three groups of soil are unique to provinces 3, 4 and 5, yet correlation is being attempted with schemes derived from province 2. Furthermore, the analytical methods being applied have to a large extent been developed for the materials of province 2 and are not necessarily applicable to materials outside that province. While it is true that D'Hoore's problem soils are mineralogically relatively featureless, in terms of gross morphology this is certainly not so; consequently, it is possible to achieve some correlation on a morphological basis with the type profiles as defined earlier for provinces 3, 4 and 5.

In conclusion, it can be said that the present model is proving satisfactory at a high categorical level for provinces 3, 4 and 5 and is, in principle, applicable to provinces 1 and 2, but its implementation awaits further investigation. Future work needs to be directed towards an evaluation of the model in terms of the kinds of basic units required for more practical applications at lower categorical levels.

Appendix

Field texture grading

Field texture grading is the most common character reported in soil descriptions. However, a great deal more than particle size is estimated in this test, as it is in fact a determination of how soil material behaves in a homogenised state at the sticky point. Sufficient soil material is taken to fit into the palm of the hand, in a state where it can readily absorb moisture. Water is added a little at a time and kneaded into the soil material until the uniform ball so produced (the bolus) just fails to stick to the fingers. The texture grade is assessed by the behaviour of the bolus when it is compressed in the hand and also when it is subjected to shearing between thumb and forefinger (ribboning). By the application of these two tests it is possible to distinguish any of seventeen texture grades (Figure A.1) and as many intermediate states as are necessary by adding plus or minus signs . It should be emphasised that there is no necessarily direct connection between field texture grade and particle size determined by mechanical analysis. There is some co-

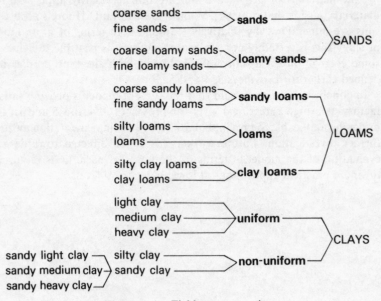

Figure A.1   Field texture grades.

incidence between the two when more sandy materials are being dealt with, but even this limited correlation decreases markedly in loams and clays.

In certain cases other important characteristics of soil material can be deduced from the behaviour of the bolus. In some soil material there is an apparent increase in the clay content the longer the bolus is manipulated, the field texture grade changing, say, from a clay loam to light to medium clay. This property is termed **subplasticity** and is associated with a fabric which is strongly resistant to mechanical breakdown. With prolonged working of the bolus such a breakdown occurs gradually, leading to an increase in the amount of finer-grained particles behaving as individuals. Other characteristics of subplastic materials are that they require continual additions of water to keep them up to the sticky point and that they tend to produce much shorter ribbons on shearing than they should, considering their clay content.

A rather different texture grade shift is caused by the presence of large amounts of fine-grained carbonates. These, on working the bolus, increase the texture grades of sands and loams, while those of clays and clay loams decrease.

Another important kind of soil material accepts water in the process of bolus formation only with the greatest reluctance. Here, the initial addition of water causes the surface to deflocculate, preventing further penetration of water. This produces a very slippery surface on an essentially unwetted bolus; it is rather like trying to grip a handful of ballbearings. This characteristic is generally associated with a considerable quantity of exchangeable sodium and/or magnesium and can be contrasted with clays dominated by exchangeable calcium or hydrogen that readily accept water; the latter form into a smooth, uniform bolus with a minimum of work.

(−2) Disintegration into finer complex entities or single grain material almost complete.

(−1) Some disintegration in which there is a definite reduction in the average size of the entities.

(0) No definite change beyond possibly a slight rounding of the entities.

(1) Coalescence has occurred to produce a considerable number of rod- or ball-shaped entities.

(2) Almost all the material has coalesced into one mass.

Figure A.2   Scale of consistence.

Consistence

Consistence (Butler 1955) is a measure of how soil material behaves when it is manipulated in its natural state. The test is performed on a 3 cm cube of soil material, which has been removed from the soil mass with as little disturbance as possible. Pressure is exerted on this material by squeezing between the palms of the hand until disruption occurs. The material is then subjected to shearing stress by slowly rubbing the palms of the hands together at the same pressure as that required to cause the initial disruption. Observations are made of the pressure required for disruption, the nature of the fragments produced by the disruption (i.e. to what extent they are pedal) and how these fragments alter with the application of shearing stress (Fig. A.2), thereby determining the degree of brittleness or plasticity. These determinations are considerably affected by the moisture status of the soil: the degree of change in reaction with change in moisture status can vary from material to material. Therefore, it is desirable for determinations to be made in both dry and moist states. It is, of course, difficult to apply this consistence test to sandy materials and to finer-textured materials in the dry state. Its use is, therefore, restricted in comparison with field texture grading, but it is of great value when it can be applied to complement and amplify observations on fabric and determinations of field texture grade.

Bibliography

Barnard, C. 1964. *Grasses and grasslands*. London: Macmillan.

Barratt, B. C. 1964. A classification of humus forms and micro-fabrics of temperate grasslands. *Soil Sci.* **15**, 342–56.

Barth, T. F. W. 1948a. The distribution of oxygen in the lithosphere. *J. Geol.* **56**, 41–9.

Barth, T. F. W. 1948b. Oxygen in rocks: a basis for petrographic calculations. *Ibid*, 50–60.

Baxter, F. P. and F. D. Hole 1967. Ant (*Formica cinera*) pedoturbation in a prairie soil. *Proc. Soil Sci. Soc. Amer.* **31**, 425–8.

Beckman, G. G., C. H. Thompson and G. D. Hubble 1974. Genesis of red and black soils on basalt on the Darling Downs, Queensland, Australia. *J. Soil Sci.* **25**, 265–81.

Beckwith, R. S. 1955. Metal complexes in soils. *Aust. J. Agric. Res.* **6**, 685–98.

Berg, R. Y. 1975. Myrmecochorous plants in Australia and their dispersal by ants. *Aust. J. Bot.* **23**, 475–508.

Berkner, L. V. and L. C. Marshall 1965. On the origin and rise of oxygen concentration in the earth's atmosphere. *J. Atmos. Sci.* **22**, 225–61.

Bernal, J. D. 1967. *The origin of life.* New York: World Publishing Co.

Birkeland, P. W. 1974. *Pedology, weathering and geomorphological research.* Oxford: Oxford University Press.

Birot, P. 1968. *The cycle of erosion in different climates.* London: Batsford.

Bloomfield, C. 1953. A study of podzolization. Part 2. *J. Soil Sci.* **4**, 17–23.

Bloomfield, C. 1954. A study of podzolization. Parts 3, 4 and 5. *Ibid*, **5**, 39–56.

Bouillon, A. 1970. Termites of the Ethiopian region. In *Biology of termites,* K. Krishna and F. M. Wessner (eds), Vol. 2, 153–280. New York and London: Academic Press.

Bowler, J. M. 1973. Clay dunes: their occurrence, formation and environmental significance. *Earth. Sci. Rev,* **9**, 315–38.

Brewer, R. 1964. *Fabric and mineral analysis of soils.* New York: John Wiley.

Butler, B. E. 1955. A system for the description of soil structure and consistence in the field. *J. Aust. Inst. Agric. Sci.* **21**, 239–49.

Carroll, D. 1962. Rainwater as a chemical agent of geologic process – *a review.* US Geol. Surv. Water Supply Paper 1535-G.

Carson, M. A. and M. J. Kirkby 1972. *Hillslope form and process.* Cambridge: Cambridge University Press.

Chenery, E. M. 1948. Aluminium in the plant world. Part 1: General survey in dicotyledons. *Kew Bull.* **2**, 173–83.

Cloud, P. 1968. Atmospheric and hydrospheric evolution on the primitive earth. *Science* **160**, 729–36.

Coleman, N. T. 1962. Decomposition of clays and the fate of aluminium. *Econ. Geol.* **57**, 1207–18.

Coulson, C. B., R. I. Davies and D. A. Lewis 1960. Polyphenols in plant, humus and soil. Parts 1 and 2. *J. Soil Sci.* **11**, 20–44.

Crocker, R. L. 1952. Soil genesis and the pedogenic factors. *Quart. Rev. Biol.* **27**, 139–68.

Crompton, E. 1960. The significance of the weathering/leaching ratio in the differentiation of major soil groups. *Trans 7th Int. Congr. Soil Sci.* **4**, 406–12.

Crompton, E. 1962. Soil formation. *Outlook on agriculture* **3**, 209–18.

Darwin, C. 1881. *The formation of vegetable mould through the action of worms, with observations on their habits.* London: John Murray.

Davies, R. I., C. B. Coulson and D. A. Lewis 1964. Polyphenols in plant, humus and soil. Parts 3 and 4. *J. Soil Sci.* **15**, 299–318.

Davis, S. N. 1964. Silica in streams and groundwater. *Amer. J. Sci.* **262**, 870–91.

De Sigmond, A. A. J. 1938. *The Principles of soil science.* London: Thomas Murby.

De Vore, G. W. 1959. The surface chemistry of feldspars as an influence on their decomposition products. *Clay and Clay Minerals* **2**, 26–41.

D'Hoore, J. L. 1964. *Explanatory monograph 1 : 5 million soil map of Africa.* CCTA Joint Project no. 11. Lagos, Nigeria.

D'Hoore, J. L. 1968. The classification of tropical soils. In *The soil resources of tropical Africa,* R. P. Moss (ed.), 7–28. Cambridge: Cambridge University Press.

Dickson, B. A. and R. L. Crocker 1953. A chronosequence of soils and vegetation near Mt Shasta, California. Part 2. *J. Soil Sci.* **4,** 142–54.

Downes, R. J. 1946. Tunnelling erosion in northeastern Victoria. *J. CSIRO* **19,** 283–92.

Duchaufour, P. 1960. *Précis de pedologie.* Paris: Masson et Cie.

Edwards, C. A. and J. K. Lofty 1972. *The biology of earthworms.* London: Chapman & Hall.

Emerson, W. W. 1959. The structure of soil colloids. *J. Soil Sci.* **10,** 235–44.

Erikson, E. 1959. The yearly circulation of chloride and sulphur in nature; meteorological, geochemical and pedological implications. Part 1. *Tellus* **11,** 375–403.

Erikson, E. 1960. Part 2. *Ibid.,* **12,** 63–109.

Evans, A. C. 1948. Studies on the relationships between earthworms and soil fertility. Part 2. Some effects of earthworms on soil structure. *Ann. App. Biol.* **35,** 1–13.

Evans, R. C. 1964. *An introduction to crystal chemistry.* Cambridge: Cambridge University Press.

Ferguson, J. A. 1954. Transformations in clay minerals in black earths and red loams of basaltic origin. *Aust. J. Agric. Res.* **5,** 98–105.

Fieldes, M. 1955. Clay mineralogy of New Zealand soils. Part 2. Allophane and related mineral colloids. *N.Z. J. Sci. Tech. B.* **37,** 336–50.

Fieldes, M. 1966. The nature of allophane in soils. Part 1. Significance of structural randomness in pedogenesis. *N.Z. J. Sci.* **9,** 599–607.

Fieldes, M. and L. D. Swindale 1954. Chemical weathering of silicates in soil formation. *N.Z. J. Sci. Tech. B* **36,** 140–54.

Fieldes, M. and N. H. Taylor 1961. Clay mineralogy of New Zealand soils. Part 5. Mineral colloids and genetic classification. *N.Z. J. Sci.* **4,** 679–706.

Flint, R. F. 1971. *Glacial and Quaternary geology.* New York: John Wiley.

Frederickson, A. F. 1951. Mechanism of weathering. *Bull. Geol. Soc. Amer.* **62,** 221–32.

Frederickson, A. F. and J. E. Cox 1953. The decomposition products of anorthite attacked by pure water at elevated temperatures and pressures. *Proc. 2nd National Conf. Clay and Clay Minerals,* 111–19.

Frederickson, A. F. and J. E. Cox 1954. Mechanism of 'solution' of quartz in pure water at elevated temperatures and pressures. *Amer. Min.* **39,** 886–901.

Giardino, J. R. 1974. When elephants destroy a valley. *Geog. Mag.* **47,** 175–81.

Gibbs, H. S. 1949. The effect of cyclic salt on coastal soils near Wellington and some regional applications. *Proc. 7th Pac. Sci. Cong.* **6,** 28.

Gilkes, R. J. 1973. The alteration products of potassium depleted oxybiotite. *Clay and Clay Minerals* **21,** 303–13.

Gilkes, R. J., R. C. Young and J. P. Quirk 1972. Oxidation of ferrous iron in biotite. *Nature, Phys. Sci.* **236,** 89–91.

Gilkes, R. J., R. C. Young and J. P. Quirk 1973. Artificial weathering of oxidized biotite: I. Potassium removal by sodium chloride and sodium tetraphenylboron solutions: II. Rates of dissolution in 0·1, 0·01, 0·001M HCl. *Proc. Soil Sci. Soc. Amer.* **37,** 25–33.

Glover, P. E., E. C. Trump and L. E. D. Wateridge 1964. Termitaria and vegetation patterns on the Loita Plains of Kenya. *J. Ecol.* **52,** 365–77.

Goldschmidt, V. M. 1937. The principles of distribution of chemical elements in minerals and rocks. *J. Chem. Soc.* **139**, 655–73.

Gorham, E. 1961. Factors influencing supply of major ions to inland waters, with special reference to the atmosphere. *Bull. Geol. Soc. Amer.* **72**, 795–840.

Greaves, T. 1962. Studies of foraging galleries and the invasion of living trees by *Coptotermes acinaciformis* and *C. bruneus* (Isoptera). *Aust. J. Zool.* **10**, 630–51.

Greenland, D. J. 1965a. Interactions between clays and organic compounds in soils. Part 1 – Mechanisms of interaction between clays and defined organic compounds. *Soils and Fertilizers* **28**, 415–25.

Greenland, D. J. 1965b. Interactions between clays and organic compounds in soils. Part 2 – Adsorption of soil organic compounds and its effect on soil properties. *Ibid*, 521–32.

Guild, W. M. McL. 1955. Earthworms and soil structure. In *Soil zoology*, D. K. McD. Kevan (ed.), 83–98. London: Butterworths.

Hallsworth, E. G. 1951. An interpretation of the soil formations found on basalt in the Richmond Tweed region of New South Wales. *Aust. J. Agric. Res.* **2**, 411–28.

Handley, W. R. C. 1954. *Mull and mor formation in relation to forest soils.* UK Forestry Commission Bulletin 23.

Hay, R. L. 1959. Origin and weathering of late Pleistocene ash deposits on St. Vincent, British West Indies. *J. Geol.* **67**, 65–87.

Hay, R. L. and B. F. Jones 1972. Weathering of basaltic tephra on the island of Hawaii. *Bull. Geol. Soc. Amer.* **83**, 317–32.

Hem, J. D. and W. H. Cropper 1959. Survey of ferrous–ferric chemical equilibria and redox potentials. USGS Water Supply Paper 1459A.

Hissink, D. J. 1938. The reclamation of the Dutch saline soils and their further weathering under humid climatic conditions of Holland. *Soil Sci.* **45**, 83–94.

Huggins, M. L. and Kuan-Han Sun 1946. Energy additivity in oxygen containing crystals and glasses. *J. Phys. Chem.* **50**, 319–28.

Hutchinson, G. E. 1943. The biogeochemistry of aluminium and certain related elements. *Quart. Rev. Biol.* **18**, 1–346.

Hutton, J. T. 1958. The chemistry of rainwater with particular reference to conditions in south eastern Australia. In *Arid Zone Research, Climatology and Microclimatology. Proceedings of the Canberra Symposium* **11**, 285–90.

Imeson, A. C. 1976. Some effects of burrowing animals on slope processes in the Luxembourg Ardennes. *Geograf. Ann.* **58A**, 115–25.

Jacks. G. V. 1953. Organic weathering. *Sci. Prog.* **41**, 301–5.

Jenny, H. 1941. *Factors of soil formation.* New York: McGraw-Hill.

Joffe, J. S. 1936. *Pedology.* New Brunswick: Rutgers University Press.

Jongerius, A. and J. Schelling 1960. Micromorphology of organic matter formed under the influence of soil organisms, especially soil fauna. *7th Int. Cong. Soil. Sci.* **2**, 702–10.

Keller, W. D. 1954. The bonding energies of some silicate minerals. *Amer. Min.* **39**, 783–93.

Laws, J. O. and D. A. Parsons 1943. The relation of raindrop size to intensity. *Trans. Amer. Geophy. Union* **24**, 452–9.

Lee, K. E. and T. G. Wood 1971. *Termites and soils.* London: Academic Press.

Lee, K. E. and T. G. Wood 1971. Physical and chemical effects on soils of some Australian termites and their pedological significance. *Pedobiologica* **2**, 376–409.

Lovering, T. S. 1959. Significance of accumulator plants in rock weathering. *Bull. Geol. Soc. Amer.* **70**, 781–800.

Lutz, H. J. and F. S. Griswold 1939. The influence of tree roots on soil morphology. *Amer. J. Sci.* **237**, 389–400.

Madge, D. S. 1969. Field and laboratory studies on the activity of two species of tropical earthworms. *Pedobiologica* **9**, 188–214.

Mattson, S., G. Sandberg and P. Terning 1944. Electro-chemistry of soil formation. VI. Atmospheric salts in relation to soil and peat formation and plant composition. *Ann. Agric. Coll. Sweden* **12**, 101–18.

Melton, D. A. 1976. The biology of aardvark. (Tubulidentata-Orycteropodidae). *Mammal Rev.* **6**, 75–88.

Meyer, J. A. 1960. Résultats agronomiques d'un essai de nivellement des termitières réalisé dans la Cuvette centrale Congolaise. *Bull. Agric. Congo Belge* **51**, 1047–59.

Milne, G. 1947. A soil reconnaissance journey through parts of Tanganyika Territory, December 1935 to February 1936. *J. Ecol.* **35**, 192–265.

Murata, K. J. 1942. The significance of internal structure in gelatinizing silicate minerals. *U.S.G.S. Bull.* **950**, 25–33.

Nicolls, K. D. and B. M. Tucker 1956. Pedology and chemistry of the basaltic soils of the Lismore district, N.S.W., *C.S.I.R.O. Aust. Soil Pub.* **7**.

Northcote, K. H. 1971. *A factual key for the recognition of Australian soils.* South Australia: Rellim.

Nye, P. H. 1954. Some soil forming processes in the humid tropics. Part 1. *J. Soil Sci.* **5**, 7–20.

Nye, P. H. 1955. Parts 2, 3 and 4. *Ibid* **6**, 51–83.

Nye, P. H. and D. J. Greenland 1960. *The soil under shifting cultivation.* Tech. Comm. 51. Harpenden: Comm. Bur. Soils.

Okamoto, G., T. Okura and K. Goto 1957. Properties of silica in water. *Geochim. Cosmochim. Acta* **12**, 123–32.

Ollier, C. D. 1969. *Weathering.* Edinburgh: Oliver and Boyd.

Panebokke, C. R. and J. P. Quirk 1957. Effect of initial water content on stability of soil aggregates in water. *Soil Sci.* **83**, 185–95.

Parle, J. N. 1963. A microbiological study of earthworm casts. *J. Gen. Microbiol.* **13**, 13–23.

Paton, T. R. 1974. Origin and terminology for gilgai in Australia. *Geoderma* **11**, 221–42.

Paton, T. R., P. B. Mitchell, D. Adamson, R. A. Buchanan, M. D. Fox and G. Bowman 1976. Speed of podzolization. *Nature* **260**, 601–2.

Paton, T. R. and M. A. J. Williams 1972. The concept of laterite. *Ann. Assoc. of Amer. Geog.* **62**, 42–56.

Pickering, R. J. 1962. Some experiments on three quartz-free silicate rocks and their contribution to an understanding of laterization. *Econ. Geol.* **57**, 1185–206.

Prescott, J. A. 1931. The soils of Australia in relation to vegetation and climate. *C.S.I.R.O. Aust. Bull.* **52**.

Price, L. W. 1971. Geomorphic effect of the Arctic ground squirrel in an alpine environment. *Geograf. Ann.* **53A**, 100–6.

Ratcliffe, F. N. and T. Greaves 1940. The subterranean foraging galleries of *Coptotermes leactus* (Frogg). *C.S.I.R.O. Aust. Jour.* **13**, 150–61.

Ratcliffe, F. N., F. J. Gay and T. Greaves 1952. *Australian termites.* Melbourne: C.S.I.R.O.

Rodin, L. E. and N. I. Bazilevich 1965. *Production and mineral cycling in terrestrial vegetation.* Edinburgh: Oliver and Boyd.

Russell, E. W. 1961. *Soil conditions and plant growth.* London: Lowe and Brydone.

Ruxton, B. P. 1966. Correlation and stratigraphy of dacitic ash fall layers in north eastern Papua. *J. Geol. Soc. Aust.* **13**, 41–67.

Salisbury, E. J. 1925. Note on the edaphic succession in some dune soils with special reference to the time factor. *J. Ecol.* **13**, 322–8.

Satchell, J. E. 1958. Earthworm biology and soil fertility. *Soil Fert.* **21**, 209–19.

Satchell, J. E. 1967. Lumbricidae. In *Soil biology*, A. Burges and F. Raw (eds), 259–322. London: Academic Press.

Shaler, N. S. 1890–91. The origin and nature of soils. *U.S.G.S. 12th Annual Report*, 213–345.

Shantz, H. L. and C. F. Marbut 1923. *The vegetation and soils of Africa.* Washington: National Geographic.

Sharpe, C. F. S. 1938. *Landslides and related phenomena.* New York: Columbia University Press.

Sherman, G. D. 1952. The genesis and morphology of the alumina-rich laterite clays. In *Problems of clay and laterite genesis*, 154–61. New York: Amer. Inst, Min. Met. Eng.

Singer, A. 1966. The mineralogy of the clay fraction from basaltic soils in the Galilee, Israel. *J. Soil Sci.* **17**, 136–47.

Slater, C. S. and H. Hopp 1947. Leaf protection in winter to worms. *Proc. Soil Sci. Soc. Amer.* **12**, 508–11.

Soil Survey Staff 1960. *Soil classification, a comprehensive system (7th Approximation).* Washington: U.S.D.A. Soil Cons. Serv.

Stace, H. C. T., G. D. Hubble, R. Brewer, K. H. Northcote, J. R. Sleeman, M. J. Mulcahy and E. G. Hallsworth 1968. *A handbook of Australian soils.* South Australia: Rellim.

Stephens, E. P. 1956. The uprooting of trees: a forest process. *Proc. Soil Sci. Soc. Amer.* **20**, 113–16.

Stoops, G. 1964. Application of some pedological methods to the analysis of termite mounds. In *Études sur les termites africains*, A. Bouillon (ed.), 379–98. Leopoldville University.

Swaby, R. J. 1950. The influence of earthworms on soil aggregation. *J. Soil Sci.* **1**, 195–7.

Sys, C. 1960. Principles of soil classification in the Belgian Congo. *Proc. 7th Int. Cong. Soil Sci.* **4**, 112–18.

Tamm, O. 1920. Bodenstudien in der Nordschwedischen Nadelwaldregion. *Medd. Stat Skogsförs* **17**, 49–300.

Taylor, H. P. and S. Epstein 1962. Relationship between $^{18}O/^{16}O$ ratios in coexisting minerals of igneous and metamorphic rocks. Part 1. *Bull. Geol. Soc. Amer.* **73**, 461–80.

Thomas, A. S. 1941. The vegetation of the Sese islands, Uganda. An illustration of edaphic factors in tropical ecology. *J. Ecol.* **29**, 330–53.

Thorp, J. 1949. Effects of certain animals that live in soils. *Sci. Monthly* **68**, 180–91.

Turner, F. J. and J. Verhoogen 1960. *Igneous and metamorphic petrology*. New York: McGraw-Hill.

Van Baren, J. 1931. Properties and constitution of a volcanic soil, built in 50 years in the East Indian Archipelago. *Comm. Geol. Inst. Agr. Univ. Wageningen* **17**.

Van Hise, C. R. 1904. *A treatise on metamorphism*. U.S.G.S. Monograph **47**.

Wallace, A. 1963. Role of chelating agents on the availability of nutrients to plants. *Proc. Soil Sci. Soc. Amer.* **27**, 176–9.

Waloff, N. and R. E. Blackith 1962. Growth and distribution of the mounds of *Lasius flavus* (Fabricius) (Hym: Formicidae) in Silwood Park, Berkshire. *J. An. Ecol.* **31**, 421–37.

Watson, J. A. L. and F. J. Gay 1970. The role of grass-eating termites in the degradation of a mulga ecosystem. *Search* **1**, 43.

Wheeler, W. M. 1910. *Ants: their structure, development and behaviour*. New York: Columbia University Press.

Wickmann, F. E. 1944. Some notes on the geochemistry of elements in sedimentary rocks. *Arkiv. für Kemi, Min., Geol.* **19B**, **2**, 1–7.

Williams, M. A. J. 1968. Termites and soil development near Brock's Creek, Northern Territory. *Aust. J. Sci.* **31**, 153–4.

Young, A. 1972. *Slopes*. Edinburgh: Oliver and Boyd.

Index